Anderson Dias Monteiro

Obtaining the Nb-Cu Composite

Anderson Dias Monteiro

Obtaining the Nb-Cu Composite

Via powder metallurgy

ScienciaScripts

Imprint

Any brand names and product names mentioned in this book are subject to trademark, brand or patent protection and are trademarks or registered trademarks of their respective holders. The use of brand names, product names, common names, trade names, product descriptions etc. even without a particular marking in this work is in no way to be construed to mean that such names may be regarded as unrestricted in respect of trademark and brand protection legislation and could thus be used by anyone.

Cover image: www.ingimage.com

This book is a translation from the original published under ISBN 978-613-9-63364-7.

Publisher:
Sciencia Scripts
is a trademark of
Dodo Books Indian Ocean Ltd. and OmniScriptum S.R.L publishing group

120 High Road, East Finchley, London, N2 9ED, United Kingdom
Str. Armeneasca 28/1, office 1, Chisinau MD-2012, Republic of Moldova, Europe
Printed at: see last page
ISBN: 978-620-7-66671-3

DEDICATORY

To my brothers Adilson and André; to my cousin Carmen; to my sister-in-law Ivanice; to my niece Ana Luisa and, especially, to my parents, Manoel and Ana.

ACKNOWLEDGEMENTS

To Professor Dr Palloma Vieira Muterlle for her guidance, support, motivation and valuable contributions during the development of this master's work.

To my great friend and mechanical engineer Daniel Amâncio Cavalcante for his help at various stages of my work.

My friend and chemical engineer Lara Jardim Grossi for helping me at various stages of my work.

To my great friend and materials laboratory technician, Alexandre. His help with the SEM analyses was of fundamental importance.

To everyone at SG9 who contributed to the preparation and production of the experiments carried out for this master's work.

To the chemistry development engineer, Rogério Marques Ribas, from Companhia Brasileira de Metalurgia e Mineração - CBMM. The initial proposal was to use niobium powders supplied by him, but this was not possible due to the unsuitable granulometry for the powder metallurgy process.

To Professor Dr Hugo Ricardo Zschommler Sandim of the Materials Engineering Department of the Lorena School of Engineering - DEMAR-EEL-USP. The metallic niobium powder he supplied was used in the production of the Nb-Cu composite, the subject of this study.

SUMMARY

The Nb-Cu composite can be used to manufacture electrical contacts, resistors, welding electrodes and more. These materials must have high electrical and thermal conductivities and heat resistance. The production of these composites using conventional technology is limited, since they involve chemical elements that have very different melting points. An alternative is to use powder metallurgy. The aim of this work is to obtain an Nb-Cu composite via powder metallurgy. To do this, niobium and copper powders were mixed in proportions of 12.5% and 25% respectively in a planetary ball mill using the following parameters: the ball/powder ratio was 3:1; the rotation speed was 250 rpm; the atmosphere was vacuum; the process controlling agent was zinc stearate; and the milling time was 2 hours and 8 hours. The apparent densities of the niobium and copper powder mixtures were then calculated. Compaction was carried out in a universal press with a force of 3.2×10^4 N. The density of the compacted materials was then determined using the geometric method. After compaction, the samples were sintered in a tube furnace under an argon gas atmosphere at a temperature of 1000°C for batch 1 samples and 1100°C for batch 2 samples, with an isotherm of 1 hour in both cases. The heating and cooling rate was 5°C/min. The density of the samples after sintering was obtained using the Archimedean method. The elementary metal powders, the mixtures of ground metal powders and the sintered bodies were analysed using scanning electron microscopy and confocal microscopy. XRD and microhardness (HV1) measurements were also carried out. From these analyses, it was observed that increasing the milling time makes the powder mixtures more homogeneous and with flatter agglomerated particles. It was also found that the apparent densities of the powder mixtures decrease with increasing milling time, while the green compacts and sintered specimens have their densities increase with increasing milling time. It was also found that the Vickers microhardness values increased with both the increase in milling time and the increase in sintering temperature.

Keywords: Niobium, Copper, Nb-Cu Composite, Powder Metallurgy, High Energy Grinding.

SUMMARY

CHAPTER 1

INTRODUCTION

The Nb-Cu composite can be used to make electrical contacts, resistors, coils for generating high magnetic fields, welding electrodes, microwave absorbers and heat sinks. These materials must have high electrical and thermal conductivities and heat resistance. The production of this composite using conventional technology is limited, since these elements have very different melting points. An alternative is powder metallurgy. In powder metallurgy, alloys based on refractory metals can be produced (MELCHIORS, 2010).

This justifies the use of powder metallurgy in this study because this technique allows precise control of the desired chemical composition of the final product, as well as the reduction or elimination of machining operations, the purity of the products obtained, the good surface finish and the ease with which the production process can be automated. These are just a few of the reasons why powder metallurgy is a source of parts for practically all branches of industry, such as aerospace and automotive (DELFORGE et al, 2007).

The choice of copper is justified by the metal's good thermal conductivity, which allows it to be used in materials whose function is to dissipate heat. In some cases, performing this function requires resistance to high temperatures, which can be achieved by using a refractory metal (MEYERS et al, 2005) - in this case, niobium. This metal is found in several countries around the world. The discovery of a deposit in Araxá (MG) made Brazil the world's largest niobium reserve (ALVARENGA, 2013).

ROTTA (2005) analysed Nb-Cu composites with the following compositions: Nb-5%Cu, Nb-10%Cu, Nb-15%Cu and Nb-20%Cu. MELCHIORS (2011) chose a composite of

Nb-20%Cu and LIMA (2015) analysed an Nb-15%Cu composite. Therefore, this work will study two composites: an Nb-12.5%Cu composite and an Nb-25%Cu composite.

It is hoped that increasing the concentration of copper in the composite, which is the subject of this study, will improve densification. According to COSTA (2002), copper plays an important role during the sintering of the W-Cu system. This has been the evidence of recent work. Copper is the main agent responsible for densification both in solid state sintering and in the presence of a liquid phase. Copper is responsible for high densification during solid-state sintering and for dispersing and homogenising the solid phase in the liquid-phase sintering stage.

The only way to obtain the properties of electrical conductivity and heat resistance is by combining a highly conductive metal such as copper with a refractory metal such as niobium (SURYANARAYANA, 1998; TORRALBA et al, 2003). These properties, however, depend on the characteristics of the starting powders, the composition and the manufacturing process, so that the distribution and shape of the refractory phase, particle size, porosity and homogeneity of the microstructure affect the mechanical, electrical and thermal properties of the composite material (LENEL et al, 1980; MEYERS et al, 2005). Thus, during processing, the behaviour of metal powders depends on the characteristics of the initial powder. In this way, the particle size and distribution, the degree of agglomeration and aggregation, the structure of the powder and the shape of the particle all affect the development and densification of the microstructure (LENEL et al, 1980; GERMAN and BOSE, 1997; LEE and REINFORTH, 1994).

The Nb-Cu system has the characteristic of being almost mutually immiscible. This means that the solubility of Cu in Nb and Nb in Cu is negligible (COSTA et al., 2008). Liquid copper also has low wettability in niobium. These characteristics do not allow homogeneous, high-density structures to be produced through sintering when conventionally prepared powders are used. However, it is possible to produce a material from an immiscible system by means of powder metallurgy (PM) (MEYERS et al, 2005; NIKOLAEV and ROZENBERG, 1972).

Powder metallurgy (PM) is a technique for obtaining metal matrix composites (CMM). When compared to materials produced through rolling and casting processes, it has several advantages. These include the low manufacturing temperature, which avoids undesirable reactions at the interface between the dispersed phase and the matrix. There are also materials that can only be produced using PM, such as Ti alloys with SiC reinforcement. Another advantage of this technique is the uniformity obtained in the distribution of the dispersed phase, which not only improves structural properties but also increases reproducibility (TORRALBA et al, 2003).

Therefore, the general objective of this study is to obtain an Nb-Cu composite via the powder metallurgy process. To achieve this, the specific objectives are:

• Produce two mixtures of Nb and Cu elemental powders containing 12.5% and 25% by mass of Cu respectively;

• Characterise the composite obtained. This characterisation will be carried out as follows: - analyse the powder mixture as well as the sintered specimens using scanning electron microscopy (SEM) and confocal microscopy; - carry out sintering temperature analysis using the phase diagram; - determine the apparent density of the powder mixtures, the density of the green compact and the density of the sintered body; - carry out X-ray diffractometry analysis; - determine the microhardness of the sintered body.

Chapter 2 of this work presents the theoretical framework. This part will give a brief history of the powder metallurgy process, its definition, the stages of the process, its main applications, the sintering that takes place in the liquid and solid phases. Some properties of niobium and copper, their history and curiosities will also be described.

Chapter 3 of this work describes the materials and methods used to obtain the Nb-12.5%Cu and Nb-25%Cu powders, sintering, density measurements and how Vickers microhardness measurements were carried out.

The results and discussions will be presented in Chapter 4. This will show the influence of milling time and increased copper concentration on the formation of agglomerates, as well as the density, microhardness, sintering and diffractograms of each composite. Chapter 5 presents the conclusions and suggestions for future work.

CHAPTER 2

THEORETICAL FRAMEWORK

2.1 BRIEF HISTORY

Although archaeological research shows that man was already producing weapons, tools and spears from iron agglomerates around 6000 years BC, it was only in the 19th century that the first steps were taken towards the development of modern powder metallurgy. Specifically, the year 1829 can be considered a real milestone in this process, as it was from then on that malleable platinum parts were produced, a material that, until then, could not be processed through normal casting, given its high melting point (1770°C) (MORO and AURAS, 2007).

The beginning of the 20th century saw the development of processes to obtain parts made from tungsten, which has a melting point of 3410°C, and molybdenum, with a melting point of 2610°C, using powder metallurgy. However, actual production only expanded after the Second World War, due to the urgent need to meet demand from the nascent car industry. Today, however, there are many verified industrial applications for parts produced using this process. The fact is that this technology, when compared to conventional metallurgy, has proved to be substantially competitive, both for technological reasons and for

for economic reasons. In this way, it is possible to emphasise that where there is a need to produce large quantities of parts, with base material that has a high melting point or complex shapes, powder metallurgy can always be used (MORO and AURAS, 2007).

2.2 DEFINITION OF THE PROCESS

It is a manufacturing method that produces parts using metal powder as a raw material. The process consists of compacting or shaping the mixture and heating it, with the aim of improving the cohesion of the internal structure. The specific example of the process consists of the temperature remaining below the melting temperature of the main element (MEYERS et al, 2005). It is a process in which maximum material savings are achieved, with few losses of raw materials. Some alloys are obtained by powder metallurgy at much lower costs than if they were made by conventional metallurgy (DELFORGE et al, 2007).

Adjusting the desired chemical composition of the final product, eliminating or reducing machining operations, the purity of the products obtained, the surface finish and the ease with which the production process can be automated are all reasons why powder metallurgy is seen as a source of parts for basically all branches of industry, such as IT, automotive, aerospace, agricultural implements and equipment, electrical and electronic equipment, textiles, among others (MORO and AURAS, 2007). Figure 1 illustrates complex parts that have been produced by powder metallurgy.

Figure 1: Parts produced through powder metallurgy
Source: adapted from MORO and AURAS, 2007

Therefore, the limitations that cannot be overcome make powder metallurgy a poor solution at times. In other situations, powder metallurgy is not the last resort. The fact is that when the part is extracted from a root, it changes the production of parts with certain geometric characteristics. The advantages are as follows: almost no shortage of raw materials, drastic adjustments to the chemical composition, dimensional tolerance, good surface finish, controlled and effective use of energy and easy handling of automation. The disadvantages, on the other hand, are: a large increase in the use of tools, limitation of the parts and impediment to welding, as well as compromised chemical and physical properties due to the porosity of the part (MORO and AURAS, 2007).

2.3 STAGES OF POWDER METALLURGY

Powder metallurgy includes the following processes: obtaining powders, characterising powders, compacting powders and sintering (MEYERS et al, 2005). Figure 2 illustrates the stages of the process:

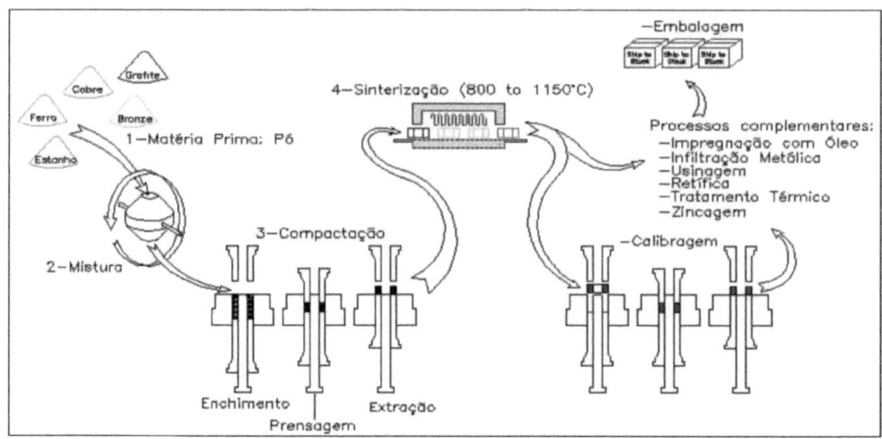

Figure 2: Powder metallurgy steps
Source: adapted from SCHAERER, 2006

1 Powder is made using specific processes;

2. the powders are mixed to make the piece homogeneous;

3. the mixture is placed in the matrix, which is pressed and extracted immediately afterwards;

4. The piece is then sent to our special labs, where it is synthesised;

5. The part goes to the complementary process to improve heat treatment (properties), calibration (tolerances), among others.

2.4 APPLICATIONS

1. Synthesised filters: these are used in a variety of industrial applications that require resistance to high temperatures and mechanical and chemical resistance (e.g. mineral oils and fuels, gas filtration, etc.). They are also used as flame arresters and noise mufflers. They are made from bronze, nickel, stainless steel, titanium and others (SCHAERER, 2006);

2. Self-lubricating bearings: The main characteristics of synthesising is to help control the porosity of the final product. This characteristic is very relevant for the production of self-lubricating bearings. The porosity contained in the bearing can be filled with oil to enable permanent lubrication between the shaft and the bearing (MELCHIORS, 2010);

3. Batteries: They use porous nickel in cadmium-nickel accumulators and batteries (SCHAERER, 2006);

4. Prostheses: Surgical implants are covered with a porous alloy (based on Co-Ti), allowing bone tissue to penetrate the pores, thus ensuring a good bond with the implant (MELCHIORS, 2010);

5. Special materials: These are alloys, such as W90-Cu-Ni, which cannot be manufactured industrially due to their high density (18 g/cm^3) and high melting point (3410°C), as well as other characteristics. They also have a high capacity to absorb radiation, highlighting their use in the nuclear sector, as well as for armouring (military sector) (SCHAERER, 2006);

6. Brake and clutch discs: These are made from an iron-copper or copper alloy to which friction agents (SiO2, SiC, A12O3) and lubricating agents (C, Pb, MoS2) are added (MELCHIORS, 2010);

7. Carbide: Carbide is important in the field of wear parts, cutting tools and rock drills. In such tools, the carbide is adapted to the cutting parts in the form of an insert. In these inserts, there is a high hardness that practically equals that of diamond, withstanding temperatures of up to 1,000°C without suffering cutting loss (SCHAERER, 2006). Figure 3 illustrates some of the products obtained using powder metallurgy.

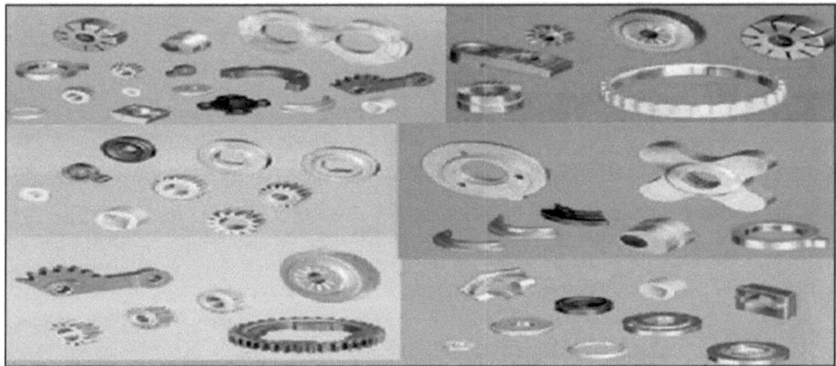

Figure 3: Gears manufactured by powder metallurgy
Source: adapted from SCHAERER, 2006

Carbide can be produced by mixing pure tungsten powder (W) with carbon powder (C) in controlled proportions to obtain the correct composition. The mixture is brought to a temperature of around 1,700°C, at which point the carbon and tungsten bond, giving rise to hard carbide particles, which are represented by tungsten carbides (WC). The cobalt (Co) added to the mixture acts as a binding metal, facilitating the dissolution of the tungsten carbide. The end result mixes the properties of the hard particle - wear resistance - with the properties of the binding metal - toughness and impact absorption (MELCHIORS, 2010);

8. electrical contacts: These are produced from Ag-WC, Ag-Ni Ag-W and pure W alloys. They   provide good thermal and electrical conductivity, preserving mechanical strength when subjected to heating, and are resistant to welding tendency when in service, with high wear resistance (SCHAERER, 2006);

9 Magnetic alloys: It is possible to guarantee small magnets and permanent magnets in

Co5Sm (Sm = samarium), polar parts, ceramic magnets in barium ferrite and magnetic cores that are widely used in radio and car speakers, engines and TVs (MELCHIORS, 2010);

10 Mechanical parts: Made from Fe-Cu-Ni, Fe, Fe-Cu-Pm[I], Fe-Cu-Ni-Mo, with or without carbon. They are used in the crankshaft, camshaft gears, bushings

homokinetic joint, forks, synchroniser rings, piston, water pump hub-pulley, valve body and guide, shock absorbers and gearbox (synchroniser rings, hubs and sleeves) etc. (SCHAERER, 2006).

2.5 POWDER PRODUCTION PROCESSES

According to Moro and Auras (2007), there are several existing processes for obtaining metal powder. Choosing the most suitable one will depend on analysing a set of material properties, as well as the characteristics you want to obtain for the powder, due to the desired application. Among the main existing methods, the authors indicate the following: *cold-stream,* mechanical methods, atomisation, electrolysis process and chemical reduction process.

The *cold-stream* method maximises the fragility of metals at low temperatures so that they can be turned into powder. Still coarse, the powder is dragged under high pressure by a stream of gas through a tube until it reaches a large chamber kept under vacuum. When it reaches the chamber, the gas expands, sharply reducing its temperature. At high speed, it collides with a target that is installed in the chamber, breaking up into smaller particles as it is relatively fragile due to the low temperature. At this point, the powder that is small enough is separated from the gaseous fraction, which is returned to the process (MORO and AURAS, 2007).

Mechanical methods (breaking and grinding) are recommended for materials that have been weakened by a previous process or are fragile. These are methods that fragment the material using mills or hammers until a certain particle size is obtained. The most commonly used mills are vibrating mills, ball mills and attrition mills (MORO and AURAS, 2007).

In atomisation, the molten metal is poured through a hole suitable for this operation, giving rise to a liquid fillet, which is attacked by jets of air (process called R-Z - *Roheisen-Zunder),* water or gas (process called ASEA-STORA, which uses nitrogen and argon, also called CSC / *Centrifugal Shot Casting* process*).* These jets lead to pulverisation of the fillet and its immediate cooling. This collects the powder, which is sieved and reduced, ready for use. The shape and size of the particles can vary due to various parameters, including fluid pressure, fillet thickness, the geometry of the spray assembly and the type of atomisation. This is the case with water atomisation, which usually produces angular and irregular particles, while air atomisation leads to spheroidal particles (MORO and AURAS, 2007).

[I] Promethius.

In the electrolysis process, the powders produced have high purity and low apparent density, with grains that have a clearly dendritic structure. After being collected from the electrolysis tanks, the slurry-like mass of powder is dried and sorted by sieving (MORO and AURAS, 2007).

The chemical reduction process, considered to be that in which a solid or gaseous reducing agent is used, is considered to be the most significant type for obtaining powders. Among the reducing agents, hydrogen and carbon stand out. Reduction with carbon can only be used for metallic elements that do not form very stable carbides, unless you want to obtain carbide powder as the end product, and not metallic powder. This is the case, for example, with tungsten carbide, which can be reduced and carbide obtained in a single treatment (MORO and AURAS, 2007).

In addition to all the methods mentioned above, we also have the thermal decomposition of carbonyls and the Hydriding-Dehydriding (HDH) process.

The thermal decomposition of carbonyls is a process in which carbonyls are obtained by reacting the metal with carbon monoxide under pressures and temperatures. Many metals react with carbon monoxide (CO) under certain conditions to form a compound called a metal carbonyl (Me (CO) x). Metal powders can then be obtained by decomposing metal carbonyls. The result of this process is very fine particles with an approximately spherical shape and high purity (JÚNIOR, 2007).

We also have the HDH process. This process consists of the embrittlement of a reactive metal with the introduction of hydrogen, using a furnace under high vacuum in the temperature range of 400-700°C with partial pressures of H2, for a period of around 4 hours (hydriding). After this stage, it is ground in a ball mill under a high purity argon atmosphere. Now, in powder form, it is degassed in a vacuum at a temperature of 650-760°C until the hydrogen is completely removed (dehydriding). FROES and EYLON (1984)

In many cases, the reaction is simple and straightforward, consisting merely of the contact of hydrogen gas (H$_2$) with the metal (M). MITKOV and BOZIC (1996). This reaction is represented by equation 1.

$$M + H_2 \leftrightharpoons MH_2 \tag{1}$$

The arrow in both directions indicates that the reaction is reversible, and the equilibrium is determined by the pressure of the hydrogen gas. If the pressure is above the equilibrium pressure, the reaction moves to the right to form the hydride; if it is below, the hydride decomposes into metal and hydrogen gas. MITKOV and BOZIC (1996).

2.6 POWDER CHARACTERISTICS

The characterisation of the powders used is of paramount importance in powder metallurgy, since their morphology, density, size, compressibility and chemical composition affect their subsequent processing (KNEWITZ, 2009).

Particle size and size distribution are of significant importance in the powder metallurgy process, being responsible for a large part of the final properties of the part obtained (ASM, 1990).

Different techniques are used to assess the size of a particle, among the most common are: sieving, light scattering, analysis by optical microscopy and analysis by scanning electron microscopy (CHIAVERINI, 2001; KNEWITZ, 2009).

The behaviour of metal powders during processing depends on the characteristics of the initial powder, such as particle size, particle size distribution, particle shape and powder structure, degree of aggregation and agglomeration; exerting a profound influence on densification and microstructure development (LENEL et al., 1980; GERMAN et al., 1985 and LEE et al., 1994).

Among the most important characteristics of metal powders are the size and shape of the individual particles. Considering the design of a sintered part, the particle size distribution is also important. The various methods used to obtain metal powders lead to products of different sizes, shapes and distribution, as well as other powder characteristics. In this context, it is essential to master the process of characterising and obtaining the powders in order to achieve a final part capable of meeting engineering requirements (MELCHIORS, 2010).

a) Geometric shape of the particles: the most common shapes are acicular, tooth-shaped, angular, fibrous, granular, porous, nodular, irregular or spherical (MELCHIORS, 2010). Figure 4 shows the geometric shapes of the particles.

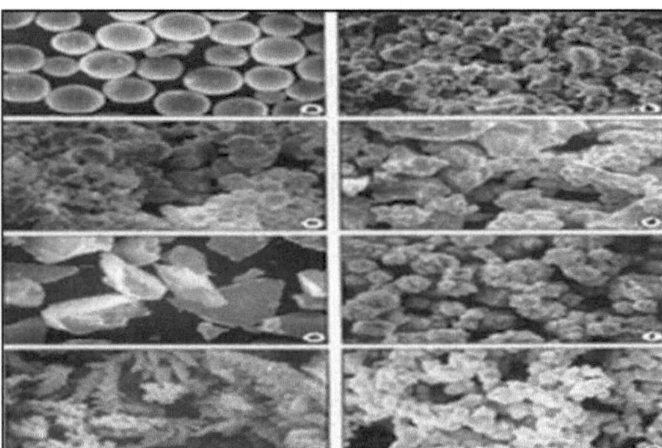

Figure 4: Morphology of metal particles
Source: adapted from KNEWITZ, 2009

b) Size: the average diameter of the particles influences certain characteristics of the final product. A larger particle, for example, provides greater compaction, while a smaller one allows for a surface with less roughness (MELCHIORS, 2010).

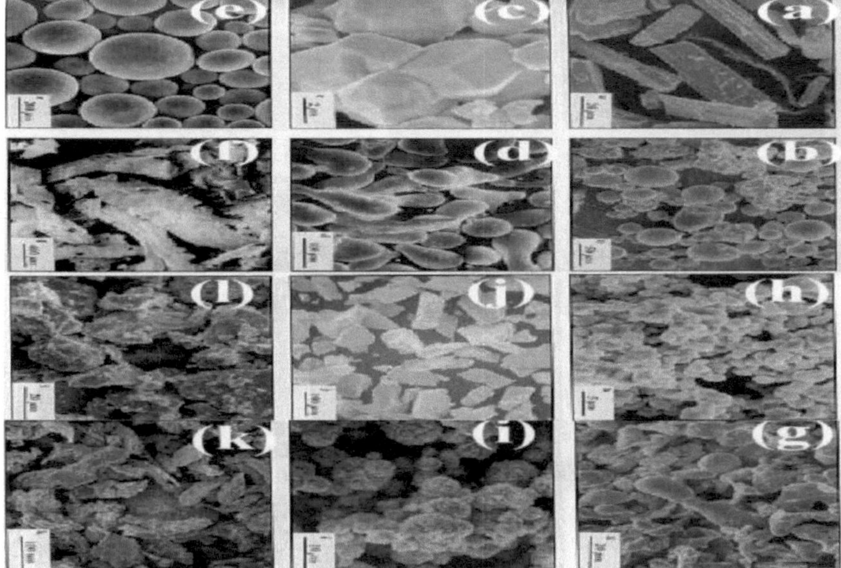

Figure 5: Characteristic powders of a material due to its manufacturing process
Source: adapted from KNEWITZ, 2009

Figure 5 shows microscopic images of various powders, characterised by the material and manufacturing process of the powder:

(a) Tellurium, ground;

(b) Alloys Fe by gas atomisation (argon);

(c) Tungsten, gas-reduced, polygonal aggregate;

(d) Tin, air atomisation, rounded with ligaments;

(e) Fe alloy, centrifugal atomisation, spherical shape;

(f) Tin, fast cooling on plate, flake format;

(g) Stainless steel, water atomisation;

(h) Electrolytic palladium, sponge;

(i) Nickel, chemical decomposition of carbon, porous;

(j) Fe metal glass;

(k) Titanium, reduction with Na and milling, irregular;

(l) Niobium hydride, grinding, irregular.

2.7 COMPACTION

The powder is placed in die cavities that are assembled in compression presses specifically manufactured for use in powder metallurgy. The powder is compressed at certain pressures, depending on the type of powder being used and the desired final characteristics of the sintered parts (CHIAVERINI, 1986). Figure 6 illustrates the components of a compaction die.

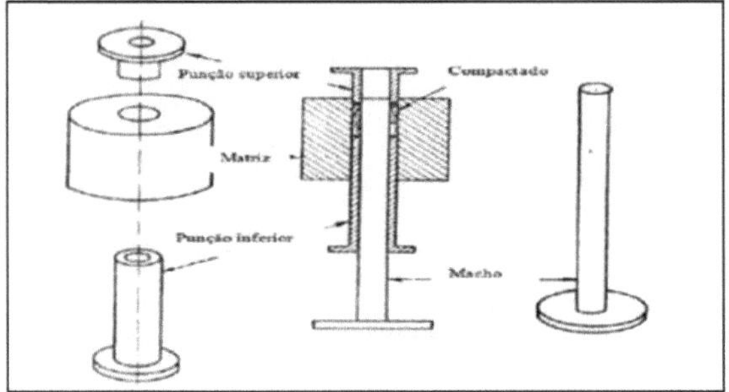

Figure 6: Simple matrix for compacting metal powders with their various components
Source: adapted from CHIAVERINI, 1986

The compression stages can be described as: filling, compression and extraction. In this stage, a predetermined amount of powder is placed in the cavity of a die mounted in a compression press, which can be mechanical or hydraulic. Compaction takes place by simultaneous movements of the upper and lower punches at room temperature. During the first movements of the punch, compaction only causes the powder to thicken, without deforming the particles or producing adhesion between them. With increased pressure, ranging from 1.6 to 9.3 t/cm^2 , the particles deform plastically, thus giving rise to a kind of "cold weld" (KNEWITZ, 2009).

Compacting produces a piece with a final shape similar to that of the part to be manufactured, which is called green compacted. This is a very fragile product that can crumble if handled improperly. It is important to emphasise the importance of the part's design, as the neutral zone can form during this stage, which is considered to be the region in which the particles have experienced the least compaction forces, resulting in a heterogeneous final part with zones of different properties. Certain types of part, with different geometries, are unfeasible precisely because of the formation of neutral zones (CHIAVERINI, 1986).

Figure 7 shows single-acting compaction, where the compaction force is exerted by only one of the two punches.

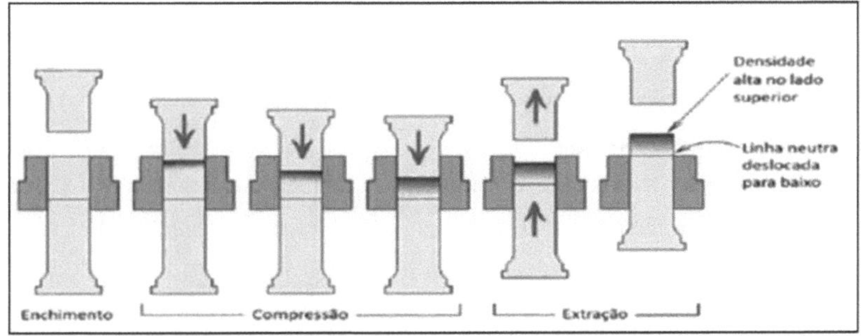

Figure 7: Compaction carried out with a single action
Source: adapted from MELCHIORS, 2010

There is the possibility of double-acting compaction where the force is exerted by both punches (KNEWITZ, 2009). Figure 8 illustrates the process.

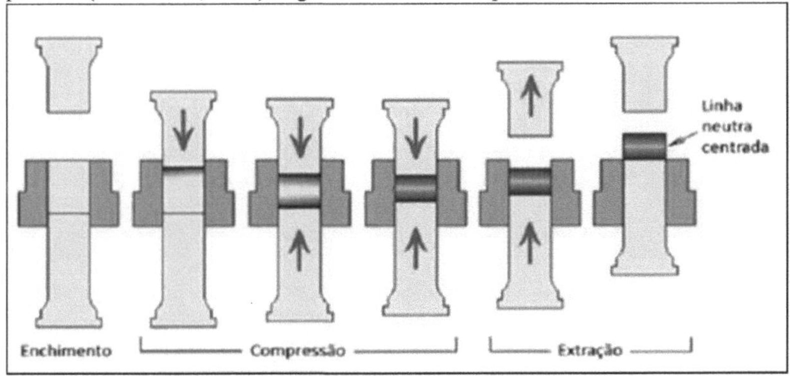

Figure 8: Compaction carried out with double action
Source: adapted from MELCHIORS, 2010

2.8 SYNTERISATION

Synthesisation can be defined as a thermally activated physical process that causes a set of particles of a given material, initially in mutual contact, to acquire mechanical resistance. Its driving force is the decrease in free energy associated with the surface of the set of particles, achieved by reducing the total surface area of the system (SILVA and JÚNIOR, 1998).

In many cases, this leads to the elimination of the empty space between the particles, resulting in a rigid and completely or partially dense body (BRITO et al, 2007).

The structural integrity of a part obtained by powder metallurgy is the result of sintering, where by heating at temperatures below the melting point the particles are joined together coherently into a solid mass. Pores are eliminated as the particles come together during high-temperature sintering, this temperature normally corresponding to 2/3 of the material's melting temperature (GERMAN and BOSE, 1997).

The green compacted material, inside or outside the matrix, is heated to high temperatures that remain below the melting temperature of the basel metal. In addition to the temperature, the speed of heating and cooling, the dwell time and the atmosphere in contact with the part are controlled. The heating time improves the cohesion mechanism
of the compacted material at a given temperature. On the other hand, temperatures close to the melting point of the metal allow the maximum cohesive strength to be achieved in a short space of time, usually seconds (BRITO et al, 2007).

Atmospheric control has four functions: it prevents or minimises chemical reactions between the green compact and atmospheric gases; it prevents oxidation; it removes existing surface and internal impurities; and it eventually provides one or more chemical elements to bind with the green compact. Sintering is normally carried out in continuous furnaces, characterised by three operating zones: preheating, temperature maintenance and cooling (GERMAN and BOSE, 1997). See:

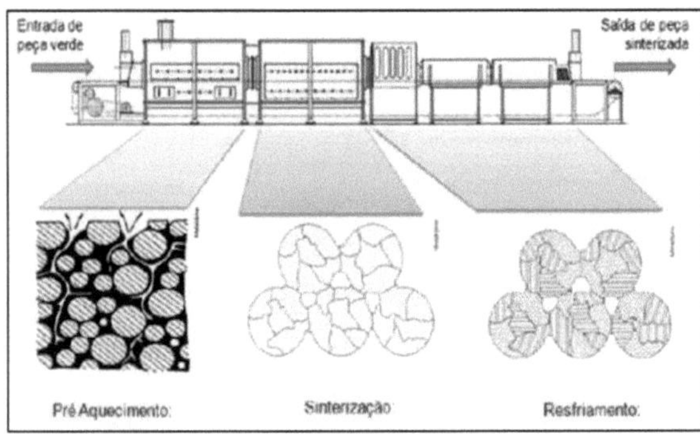

Figure 9: Stages of the conventional sintering process
Source: adapted from GERMAN and BOSE, 1997

During sintering, chemical and physical reactions take place between the particles, reducing and in some cases even eliminating the existing porosity in the green compact. The shrinkage of the green compact compared to the final piece is up to 40 per cent of the volume, a linear reduction of around 16 per cent. The phenomena that occur in sintering are as follows: initial bonding between particles; growth of the bond; closing of the channels that connect the pores; rounding of the pores; densification or contraction of the pores; eventual growth of the pores (KNEWITZ, 2009).

According to the physical state of the phases in the system, sintering is classified into: solid phase sintering and liquid phase sintering (KNEWITZ, 2009).

2.9.1 SOLID PHASE SINTERING

Solid phase sintering takes place at a temperature where none of the elements in the system reach melting point. It is carried out using material transport, such as atomic diffusion. In order to promote greater driving force, solid phase sintering sometimes involves the addition of reactive elements that alter the balance between the surface energy of the particles and the grain boundary energy, thus favouring sintering (COSTA et al, 2004).

According to Knewitz (2009), it is divided into three stages:

- the initial stage, characterised by the formation of grain contours in the area of contact between particles or the formation and growth of necks between particles, from the contacts established during the compaction process, leading up to the moment when they begin to interfere;

- intermediate stage, where there is a great densification of the compact due to the decrease in the dimensions of the interconnected pores;

- a third stage, where there is isolation of the pores in the region of the grain boundaries and gradual elimination of porosity by diffusion of vacancies from the pores along the grain boundaries. A slight densification of the structure is observed at this stage.

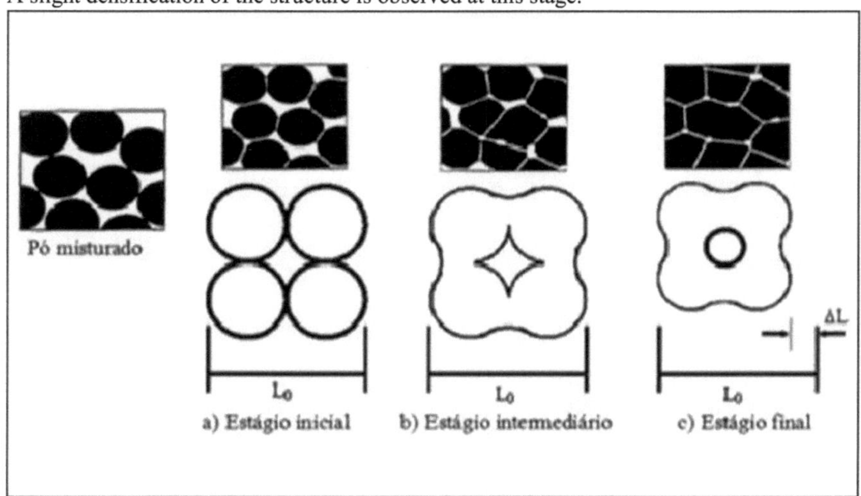

Figure 10: Representative diagram of the solid phase sintering stages
Source: adapted from GOMES, 1995 and LEE and REINFORTH, 1994

2.9.2 LIQUID PHASE SINTERING

Liquid phase synthesis is characterised by the appearance of a liquid phase at a certain sintering temperature. In this type of process, the system consists of at least two components, one of which is a liquid phase, the result of the fusion of one of the components, or the result of a reaction between them. The presence of the liquid is responsible for the speed and percentage of densification of the structure (GERMAN and BOSE, 1997).

It is also divided, according to Lee and Reinforth (1994), into three stages:

- the first stage, rearrangement or liquid flow, is marked by the spreading of the newly formed liquid around the solid particles, which leads to the rearrangement of these particles and the densification of the structure;

- the second stage is solution-reprecipitation, which will only occur if the solid phase is soluble in the liquid. When this happens, a fraction of the particles of the solid phase are dissolved by the liquid and diffuse into it, subsequently precipitating onto other solid particles in energetically more favourable locations; and

- the solid state sintering stage only takes place if the structure is not yet completely dense, and consists of neck growth between the solid parts that are in contact. At this stage, the pores close and the structure shrinks.

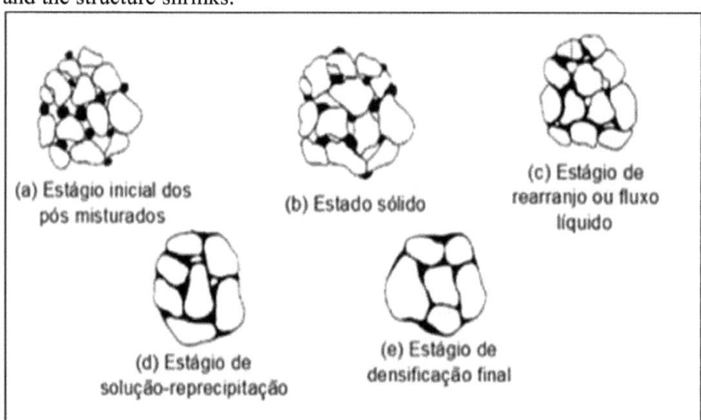

Figure 11: Representative diagram of the liquid-phase sintering stages
Source: adapted from GERMAN and BOSE, 1997

2.10 COMPLEMENTARY OPERATIONS

The production of parts by powder metallurgy does not always end with the sintering operation. Complementary operations are often applied with the following objectives: to give the parts a better finish; greater dimensional precision; better density, hardness and mechanical resistance; etc (MELCHIORS, 2010).

Calibration: during sintering, parts can undergo unexpected changes in dimensions and even warp. To correct the defects, calibration is used, which is plastic deformation by applying pressure in specific moulds. The result is better dimensional accuracy (COSTA et al., 2008).

Recompression: a new compression after sintering increases the density and improves the mechanical properties of the material. The efforts involved are much greater than in calibration and can only be applied to certain types of material. For example, carbide inserts used as machining tools cannot be recompressed. If deformations occur, they must be lapped or ground (MELCHIORS, 2010).

Heat treatments: sintered parts can be subjected to conventional heat treatments to improve their mechanical properties. In surface treatments (carburizing and nitriding), density is an important factor due to the diffusion of gases through the pores (the higher the density, the lower the porosity) (COSTA et al., 2008).

Machining: as in casting, many sintered parts undergo subsequent machining to achieve the designed configuration that cannot be made, such as holes, bleeds, threads, etc (MELCHIORS, 2010).

Infiltration: this is the process of closing the pores (totally or partially) of a sintered part with a low or medium density (5.6 to 6.8 g/cm^3) with a metal or alloy with a lower melting point. The infiltration of the liquid metal occurs due to the capillarity effect (molecular attraction), and aims to improve mechanical properties, corrosion resistance, and also as a pre-treatment for surface finishing, such as cremation, nickel plating and galvanisation (COSTA et al., 2008).

Impregnation: this consists of impregnating substances such as oils, greases and waterproofing to prevent oxidation. It is done in a hot bath, partial bath or vacuum (COSTA et al., 2008).

2.11 production of a nb-cu composite using the powder metallurgy technique

In a study on the properties of Cu-Nb alloys, Nikolaev and Rozenberg (1972), specifically the effect of niobium (up to 1.5%) on the mechanical properties, heat resistance and electrical resistivity of copper, concluded the following: the properties of Cu-Nb alloys do not depend on heat treatment conditions, which distinguishes the alloys of this system from most electrical and thermal conductors; with increasing concentrations of niobium, the hardness of the alloys increases continuously, exceeding that of copper by a factor of 2-3. The electrical conductivity also decreases appreciably.

Melchiors (2010), in an experiment carried out to obtain and characterise Nb-Cu composites by high-energy milling with vacuum furnace sintering, obtained MAE of Nb and Cu with Nb-20%Cu composition for 60 hours, producing composite particles with both crystalline phases. As Nb is harder than Cu, they found that the particles inserted into the Cu particles were deformed and fractured, while the Nb particles were also deformed.

Also according to the author, the composite particles do not resemble the original particles in shape and size, and he concluded that since the composite particles are smaller than the original Cu particles at the end of milling, there was probably fragmentation during milling due to the cold work caused by successive collisions, which harden the Cu particles, making them fragile. The author also found that during sintering, part of the Cu left the composite particles before melting, creating a layer around the particles and helping them to sinter. Later, just after melting, they found that an additional fraction of Cu had left the composite particles, filling the pores (MELCHIORS, 2010).

Lima (2015) studied a 15% Nb-Cu composite obtained by high-energy milling with liquid-phase sintering. Among the conclusions they reached, the following stand out: the density of the Nb-Cu powder structures tends to decrease as the milling time increases; the increase in milling time is directly proportional to the incompressibility of the powders and the hardening of the particles, thus making it difficult to compact the powders and consolidate them during sintering; the highest microhardness values were obtained for the samples sintered with the longest milling time and the highest temperature (1200°C); the samples sintered with the highest temperature of 1200°C showed higher electrical conductivity values when compared to the samples.

sintered at 1100°C; the density of the synthesised bodies increases at lower milling times and speeds.

Composite powders are obtained through many processes, for example: mixing, which is carried out by mixers and occurs without breaking the particles; milling, where the powders are crushed and mixed in a conventional ball mill (GOMES, 1995), or high-energy milling, which uses high milling energy to form powders made up of composite particles (LEE; REINFORTH, 1994; COSTA et al., 2008).

High-energy milling generates fine particles through the processes of plastic deformation and particle fracture (MEYERS et al, 2005). It involves constant collisions between the balls, powder and container walls, deformation, cold welding, fracture and cold re-welding of the powder particles. It produces deformation and fracture, thus defining the dispersion of the components, the homogenisation of the phases and the final microstructure of the powder. The nature of the processes depends on the mechanical behaviour of the powder components, their equilibrium phase and the state of tension during grinding (SURYANARAYANA, 1998).

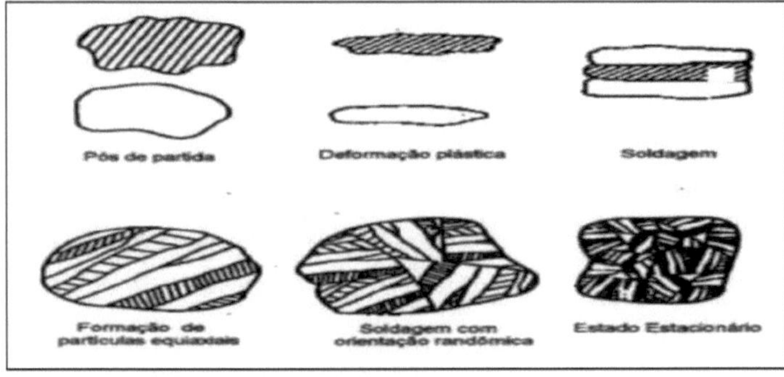

According to Melchiors (2010), MAE can be carried out with three different categories of metal powder component alloys: ductile-ductile, ductile-fragile and brittle-fragile.

2.12 DEFORMATION OF DUCTILE-FRAGILE COMPONENTS

In the initial stage of grinding, the ductile particles are flattened and the fragile constituents or intermetallic particles are fragmented by continuous collisions.

between the spheres and the powder. The fragmented fragile particles are incorporated into the ductile particles (MELCHIORS, 2010).

The bonding of ductile-fragile components during EAM requires the fragmentation of fragile particles to enable a reduction in the diffusion range and reasonable solubility in the ductile matrix components (SURYANARAYANA, 1998).

2.13 SUED BY MUM

Immiscible systems such as Nb-Ag, Nb-Cu, W-Cu, Ta-Cu, WC-Cu, W-Ag, WC-Ag, Mo-Cu, Mo-Ag, Al-In, Pb-Bi, Cu-Cr and Cu-Pb-Sn are difficult to densify by sintering due to their mutual insolubility and their high contact angle, reducing the wettability of the liquid in the solid phase (MELCHIORS, 2010).

Copper and niobium are two elements that do not mix under equilibrium conditions. In the Nb-Cu system, the three Hume-Rothery rules are violated. Firstly, copper has a face-centred cubic (CFC) crystal structure and niobium a body-centred cubic (CCC). The electronegativity values are not very close either (Nb = 1.6 and Cu = 1.9, according to the Linus Pauling scale). The difference in atomic radius between Nb and Cu is large, at over 36% (COSTA et al., 2008).Figure 13 shows the phase equilibrium diagram of the Nb-Cu system. It shows that Nb and Cu are almost completely immiscible until their melting point is reached.

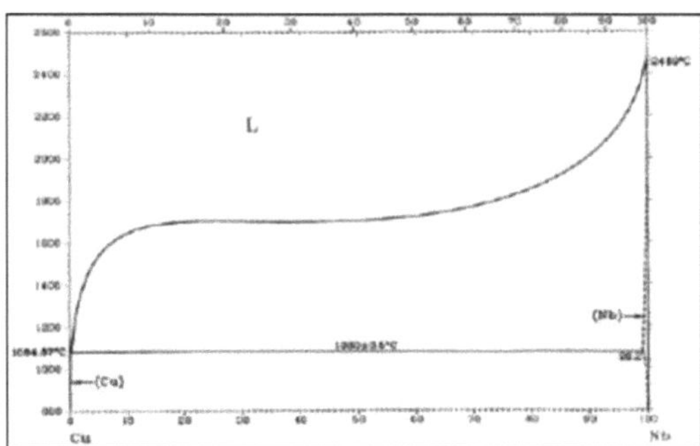

Figure 13: Phase diagram of the Nb-Cu system
Source: adapted from BAKER, 1992

2.14 NIÓBIO

Niobium is part of a group of metallic elements known as refractory metals, included with tungsten (W), tantalum (Ta), molybdenum (Mo) and rhenium (Re). They are highly resistant to heat, i.e. high temperatures, wear and corrosion. These properties make them desirable in applications for metals in the electrical and electronics industry. Important applications for refractory metals are as alloying elements in steels, forming very hard carbides, filaments for lighting, welding electrodes, lubricants, catalysts, heat sinks, electrical contacts, parts of nuclear power plant systems, among others (CHIAVERINI, 1986).

Niobium is found in the form of oxides, sometimes together with tantalum. The main niobium ores are pyrochlore [(Na, Ca)2Nb2O6F] and columbite [(Fe, Mn)(Nb, Ta)2O6] (FOGGIATO and LIMA, 2004).

Over the last 15 years, Niobium has been considered and applied in more and more chemical processes due to its excellent resistance to corrosion in many environments. Important chemical applications for Niobium include hydrochloric acid, nitric acid, chromic acid, sulphuric acid, salts, liquid metals and organic acids. Currently, the use of Niobium as an alloying element in steels, superalloys and non-ferrous alloys accounts for approximately 98% of production, as shown in Figure 2. Consumption of niobium and niobium-based alloys accounts for the remaining 5%. Niobium is in third place in the production of refractory metals, behind Molybdenum and Tungsten (MORO and AURAS, 2007).

In its natural form, niobium can be found as oxides and almost always together with tantalum. Its main ores are columbite and pyrochlore. Canada and Brazil are countries with large pyrochlore mineral reserves (MELCHIORS, 2010).

Approximately 98 per cent of the world's niobium reserves are found in Brazil, in tantalite and columbite reserves. As a result, it has become relatively rare in the rest of the world. There are several small deposits of niobium oxide found in various parts of the world, obtained as a by-product of mining other metals. In Brazil, niobium is commercially exploited by CBMM in Araxá, Minas Gerais. Large deposits of columbite are also found in Nigeria and Congo (CERNIAK, 2011).

In Rio Grande do Norte, there are columbite reserves in the municipalities of Parelhas, Carnaúba dos Dantas, Currais Novos, Equador, Jardim do Seridó, Acari, São Tomé and Santa Cruz (CERNIAK, 2011).

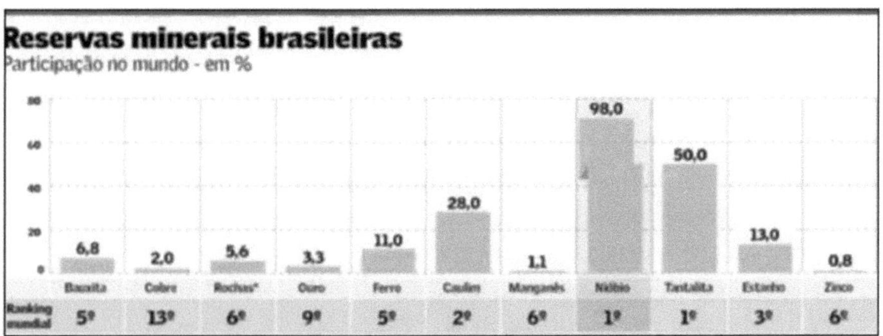

Figure 14: Brazilian mineral reserves
Source: CERNIAK, 2011

Niobium is a grey ductile metal that starts to oxidise at 200°C and must be worked in a protective atmosphere. It has superconductivity characteristics at 263.9°C (9.25K), the highest critical temperature for a pure element. It is a refractory metal with a melting temperature of 2477°C and a density of 8.57 g/cm^3 (MORO and AURAS, 2007). Other physicochemical properties of niobium are as follows:

Atomic number: 41;

Atomic weight: 92.906;

Crystal structure: CCC;

Atomic radius: 198 pm;

Electronegativity: 1.6 (Pauling scale);

Electrical conductivity: 6.579 106 /Ωm.

2.15 COVER

It is a metal with a discreetly yellowish red colour and a slightly opaque lustre with a pleasant appearance. Located in group I-B of the periodic table, it has
atomic number 29, melting point 1038°C, atomic mass 63.55 g mol^{-23} and boiling point 2927°C. It is a malleable, soft and ductile metal. Its chemical symbol is Cu, whose

origin dates back to the Latin *cuprum,* alluding to the island of Cyprus, where it is believed to have first been found (CERNIAK, 2011).

It is found in the oxidised zones of copper deposits, in association with malachite Cu_2 CO_3 $(OH)_2$, cuprite Cu_2 O and azurite Cu_3 $(CO)_{32}$ $(OH)_2$. It can also be found in double sulphide of iron and copper and in chalcopyrite $(CuFeS_2)$ (ANDRADE et al., 1997). It is one of the oldest metals in existence, given the historical period of the Bronze Age in the Neolithic period, from which copper was widely used. It is thought that copper mining began around 5,000 years ago (CERNIAK, 2011).

There was a time when copper was considered extremely rare, which caused its value to skyrocket. Soon afterwards, however, it began to be found more frequently, which forced a reduction in its cost. With the discovery of its great properties for conducting heat and electricity, copper became commercially valuable and industrialised (CERNIAK, 2011).

When copper comes into prolonged contact with atmospheric air, it oxidises and a toxic film is formed on its surface from a mixture of hydroxides, oxides and carbonates, which is green in colour and usually called azinhavre (MORO and AURAS, 2007).

Other physicochemical properties are as follows:

Atomic number: 29;

Atomic weight: 63.546;

Crystal structure: CFC;

Atomic radius: 145 pm;

Electronegativity: 1.9 (Pauling scale);

Electrical conductivity: 5.814 107 /(Ωm).

CHAPTER 3

EXPERIMENTAL PROCEDURE

3.1 MATERIALS

To produce the Nb-Cu composite, niobium, supplied by DEMAR-EEL-USP in Lorena-SP, was used as the metal matrix. In order to use it, the material was subjected to hydration followed by grinding and dehydration. This process is known as HDH. The copper powder was supplied by Metalpó Indústria e Comércio Ltda. and obtained by the carbonyl process.

3.2 MILLING

The Nb-Cu composite was produced by mixing elemental Nb and Cu powders containing 12.5% and 25% by mass of Cu respectively. This mixture was made in a planetary ball mill. Figure 15 shows this type of mill.

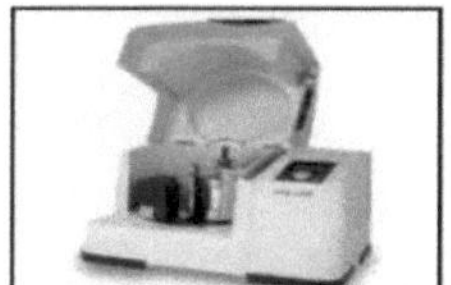

Figure 15: Planetary ball mill
Source: own elaboration

The grinding process took into account the mill's rotation speed, the grinding atmosphere, the ball-to-powder ratio used, the diameter of the balls, the process control agent and the grinding time. These parameters are listed in Table 1.

Table 1: Grinding parameters for Nb and Cu powders

Rotational speed	250 RPM
Grinding atmosphere	Vacuum
Ball to powder mass ratio	3/1
Ball diameter	10 mm
Process control agent	Zinc stearate
Grinding time	from 2 hours to 8 hours

Source: own elaboration

3.3 COMPACTION

After milling the powders, 8 (eight) specimens were produced using a compaction process. The specimens were produced in a rigid mould under uniaxial pressure of 400 MPa. These specimens are cylindrical in shape and have a diameter of 12.75 mm and a thickness of lO mm. The press and compaction matrix are shown in figure 16.

Figure 16: Press and compaction matrix
Source: own elaboration

The 8 (eight) specimens produced were identified as shown in Table 2:

Table 2: Identification of the specimens

LOT	SAMPLE IDENTIFICATION	% Cu (p.p)	% Nb (p.p)	MILLING TIME (h)	SINTERING TEMPERATURE
	L1AM1	12,5	87,5	2	
	L1AM2	25	75	2	
LOT (Ll)	L1AM3	12,5	87,5	8	1000°C
	L1AM4	25	75	8	
	L2AM1	12,5	87,5	2	
	L2AM2	25	75	2	
LOT 2 (L2)	L2AM3	12,5	87,5	8	uoo°c
	L2AM4	25	75	8	

Source: own elaboration

3.4 SYNTERISATION

To carry out the synthesis, the samples were placed in a crucible and taken to a tube furnace in an argon atmosphere, as it is an inert gas, in order to avoid the formation of oxides. The temperature was increased at a rate of 5 degrees per minute, from room temperature to 450 degrees, remaining at this temperature for 1 hour, as the zinc stearate lubricant used for grinding and compacting had to be removed from the sample. After this period, the temperature of the batch 1 samples was increased until it reached 1000°C, while the temperature of the batch 2 samples was increased until it reached 1100°C. Both sintering temperatures were maintained for 1 hour.

Both batch 1 and batch 2 samples were cooled at a rate of 5°C/minute. Figure 17 illustrates how this process will be carried out.

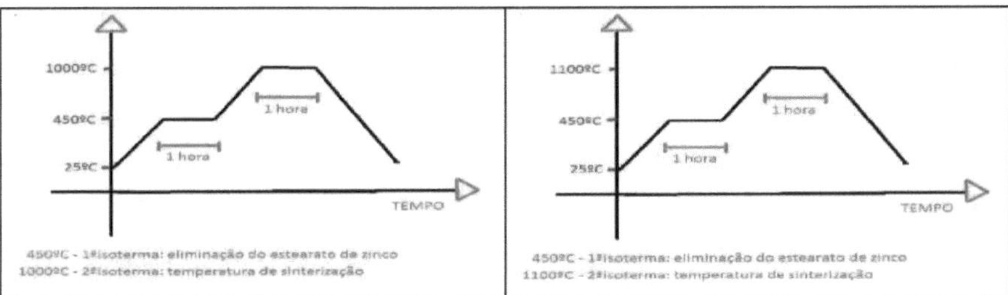

Figure 17: Demonstration of the sintering process
Source: own elaboration

3.5 DENSITY DETERMINATION

The information regarding the density of the samples was defined as follows:

1-Calculation of the bulk density of the ground powder mixture. For the bulk density, a small 2 mL graduated cylindrical container was filled with the powders from the Nb-Cu mixtures. The container containing the composite powders was tapped 10 times against a table top. When the powders decreased in volume to less than 1 mL, more powder was added to the cylindrical container. The 10 strokes of the cylinder against the table top were continued until the measured volume was equal to 1 mL. The mass of the volume occupied by the composite powder mixtures (1 mL) was weighed on an analytical balance. In this way, the ratio between the mass of the powder compressed by tapping the container against the table top and its respective volume was calculated.

2-Density calculation using the geometric method for compacted samples. After the specimens had been compacted, the diameter of the specimens and their respective height were measured using a caliper. These values were used to calculate the volume. Next, the mass of the specimens was measured on an analytical balance. In this way, the mass of the specimen and its respective volume were related, thus obtaining the density.

3-Calculation of the density of the sintered samples using the Archimedes method. The Archimedean technique is used to check the degree of densification of the composites after sintering. two sintering temperatures to which the specimens were subjected. To obtain the density using the Archimedean method, it is necessary to obtain the dry and immersed mass values. To obtain the dry mass values, the specimens were weighed on an analytical balance. To obtain the immersed mass values, it is necessary to ensure that the pores are adequately filled with water. The samples were placed in a container with water at room temperature for 1 hour. The samples were then placed in a device that kept them suspended in a container of water that remained on the analytical balance while the measurement was taken

29

To find the theoretical density of the Nb-12.5%Cu composite and the Nb- 25%Cu composite, equation 2 was used.

$$\rho T = \frac{1}{\left(\frac{\rho\%Nb}{\rho Nb}\right) + \left(\frac{\rho\%Cu}{\rho Cu}\right)} \qquad (2)$$

Where:

ρT = *theoretical density of the Nb-Cu composite*

$\rho\%Nb$ = *mass percentage of niobium in the composite*

$\rho\%Cu$ = *mass percentage of copper in the composite*

ρNb = *theoretical density of niobium*

ρCu = *theoretical density of copper*

By applying the equation, the theoretical density of the Nb-12.5%Cu composites was found to be 8.551 g/cm^3 , and Nb-25%Cu, which is 8.603 g/cm^3 . With these values for the density of the composites, it was possible to assess the evolution of the densification of the sintered material.

3.6 MICROSTRUCTURAL CHARACTERISATION AND COMPOSITION DETERMINATION

Both the copper and niobium powders were analysed using the SEM (Scanning Electron Microscope) in order to check the shape and size of the powders.

respective metallic grains. SEM was carried out on the powders that had been ground in MAE, and finally, SEM was carried out on the sintered samples. Qualitative and semi-quantitative analysis of the powders and sintered specimens was carried out using the EDS detector installed inside the microscope. Five points on the specimen were used for this analysis. The values obtained were averaged, along with the standard deviation.

3.7 CHARACTERISATION OF THE CRYSTAL STRUCTURE

X-ray diffraction (XRD) of the sintered samples was carried out using the following radiation sources

Ka of the Cu element (30 kV and 30 mA), angular range (in 2θ) used from 10° to 80° with a scan speed of 2°/min.

Through this analysis it will be possible to identify the possible phases that could be formed. And also to analyse possible changes in the crystalline structure of the composite material, as well as the presence of contaminants from the wear of the balls and the grinding container. These analyses will be carried out using diffractograms.

3.8 VICKERS MICROHARDNESS

The Vickers microhardness test was carried out using the 8 cylindrical samples. A load of 1000 gF (1 N) was applied to each sample for 10 seconds using a digital microhardness tester. This procedure was carried out five times on each specimen. The mean and standard deviation were then calculated.

Firstly, the sample was prepared by polishing it with the polishing cloth. 220, 400, 600, 800, 1200 and 2500 grit sandpaper. After using the sandpaper, alumina was used to completely polish the surface. All the stages of the experimental procedure are shown in the flowchart in figure 18.

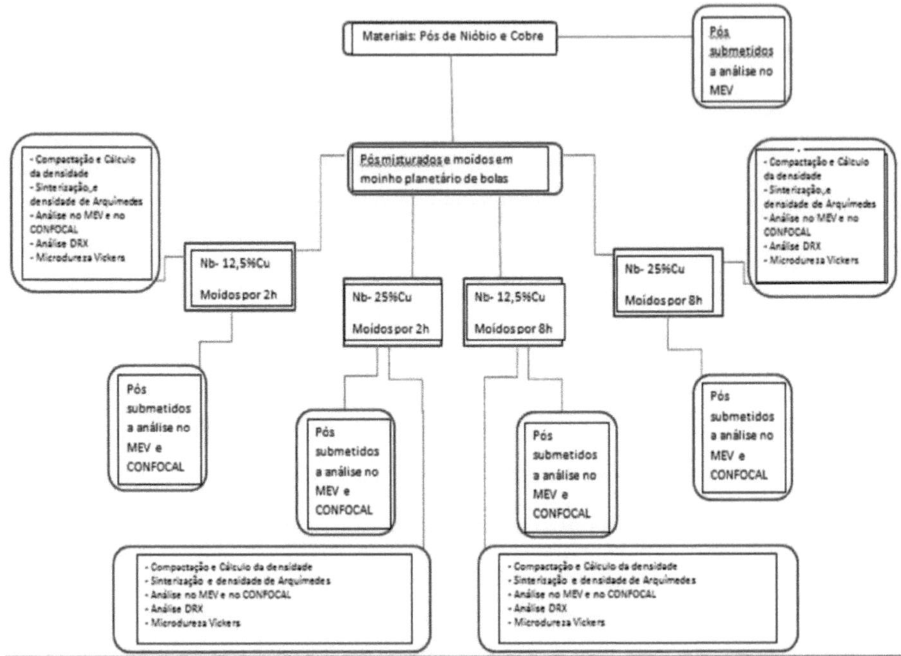

Figure 18: Flowchart of the experimental procedure
Source: own elaboration

CHAPTER 4

RESULTS AND DISCUSSIONS

4.1 MICROSTRUCTURE

In the experimental process, the analyses carried out using the SEM made it possible to obtain images of the elemental copper powders (Figure 19). It was possible to see that the particles formed had a spherical and porous morphology, which led to the conclusion that the copper powders were obtained by the carbonyl process.

Figure 19: SEM images of the copper powder
Source: own elaboration

Images of elemental niobium powder are also reproduced in this study (Figure 20). Niobium has particles with an angular morphology, typical of powders processed using the HDH (hydride-hydride) technique. The hydriding process causes niobium, which is naturally ductile, to become brittle, making it easier to break the particles when the element is exposed to grinding. This makes it possible to obtain the material in powder form.

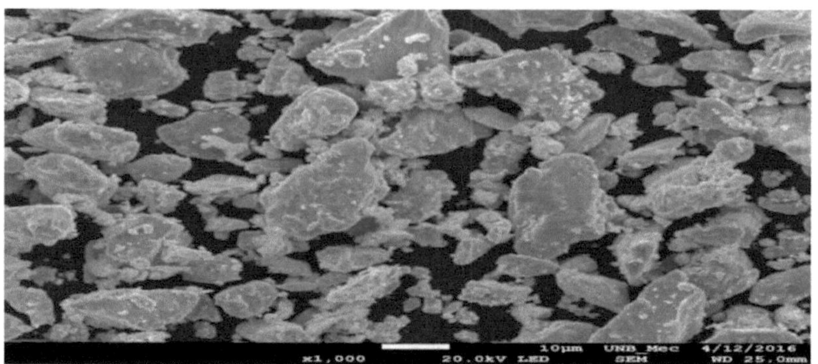
Figure 20: SEM images of niobium powder
Source: own elaboration

After mixing the copper and niobium powders into a composition of Nb-12.5%Cu, with subsequent milling for two (2) hours in a planetary ball mill, the morphology and grain size of the resulting material was analysed using a Scanning Electron Microscope (SEM) (Figure 21).

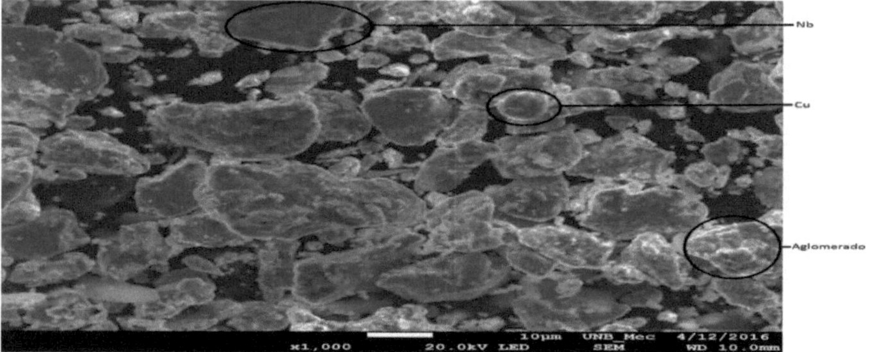

Figure 21: Images of the Nb-12.5%Cu mixture (milled for 2 hours)
Source: own elaboration

The images show that some of the particles were welded together, forming agglomerates. There was also an apparent reduction in the particle size of the elementary powders. It was also observed that some of the particles in the elementary powders did not interact. The appearance of the material obtained is heterogeneous.

Now taking a composition of Nb-25%Cu, milled for 2 (two) hours in a planetary ball mill, we also analysed the morphology and grain size of the resulting material using SEM (Figure 22). It can be seen that this composition resulted in greater welding of the particles, with the formation of more agglomerated material compared to that obtained with the Nb12.5%Cu composition (Figure 21). The particles of the elemental powders experienced an apparent reduction. The 2 (two) hour milling time does not promote complete interaction between the particles of the elemental powders, i.e. particles of the copper and niobium powders were observed. The appearance of the material obtained is also heterogeneous.

Figure 22: Images of the Nb-25%Cu mixture (milled for 2 hours)
Source: own elaboration

In grinding, deformation, cold welding and fracture are caused by successive collisions between the particles of the two phases and by the grinding bodies themselves. It can be seen that the efficiency of material fragmentation increases with increasing grinding time.

Now taking a composition of Nb-12.5%Cu, milled for 8 (eight) hours in a planetary ball mill, it was realised that the agglomerates formed had an apparent reduction in size. Plastic deformations appeared in the particles and flattened structures began to form. There was an apparent reduction in the size of the particles as the grinding time increased, because the agglomerates formed by the welding of the particles were colliding with the grinding bodies themselves and undergoing plastic deformation. The successive collisions of the balls with the agglomerates cause the particles to harden and thus fracture. Figure 23 shows this evidence.

Figure 23: Images of the Nb-12.5%Cu mixture (milled for 8 hours)
Source: own elaboration

Figure 24 shows the mixing of the copper and niobium powders into an Nb-25%Cu composition, with subsequent grinding for 8 (eight) hours in a planetary ball mill. The grinding time of 8 (eight) hours started the process of forming flattened structures. Grinding times of eight hours resulted in a material with a heterogeneous appearance.

Figure 24: Images of the Nb-25%Cu mixture (milled for 8 hours)
Source: own elaboration

4.2 SINTERING

Figure 25 shows the sintering at temperatures of 1000°C and 1100°C of the Nb-12.5%Cu and Nb-25%Cu powder samples ground for two hours. The micrographs of the powder structures show regions with large concentrations of copper (pink phase) between regions of niobium (light phase). This is a typical structure for powders prepared using the conventional powder mixing technique. The images show that the two-hour milling time gives the composite material a heterogeneous appearance.

Figure 25: Tmaeens of sintered Nb-12.5%Cu and Nb-25%Cu milled for two hours

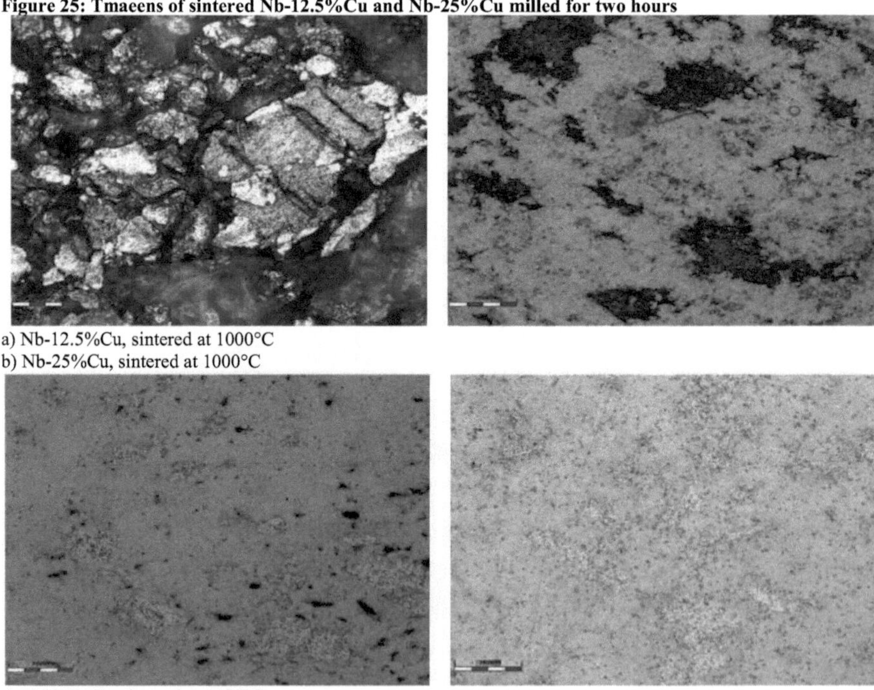

a) Nb-12.5%Cu, sintered at 1000°C
b) Nb-25%Cu, sintered at 1000°C

c) Nbl2.5%Cu, sintered at 1100°C

d) Nb-25%Cu, sintered at 1100°C
Source: own elaboration

It was observed that temperature influences sintering. Figure 26 shows that the Nb-12.5%Cu and Nb-25%Cu powders, milled for eight hours and sintered at 1000°C and 1100°C, result in a structure with a slightly higher degree of homogeneity than the samples that were milled for two hours. There are still many regions of high copper concentrations among the niobium matrix. This makes the sintered material appear heterogeneous. It has been shown that increasing the milling time and temperature increases the quality of the composite obtained. Sintering at 1100°C is more efficient.

Figure 26 - Images of sintered Nb-12.5%Cu and Nb-25%Cu milled for eight hours

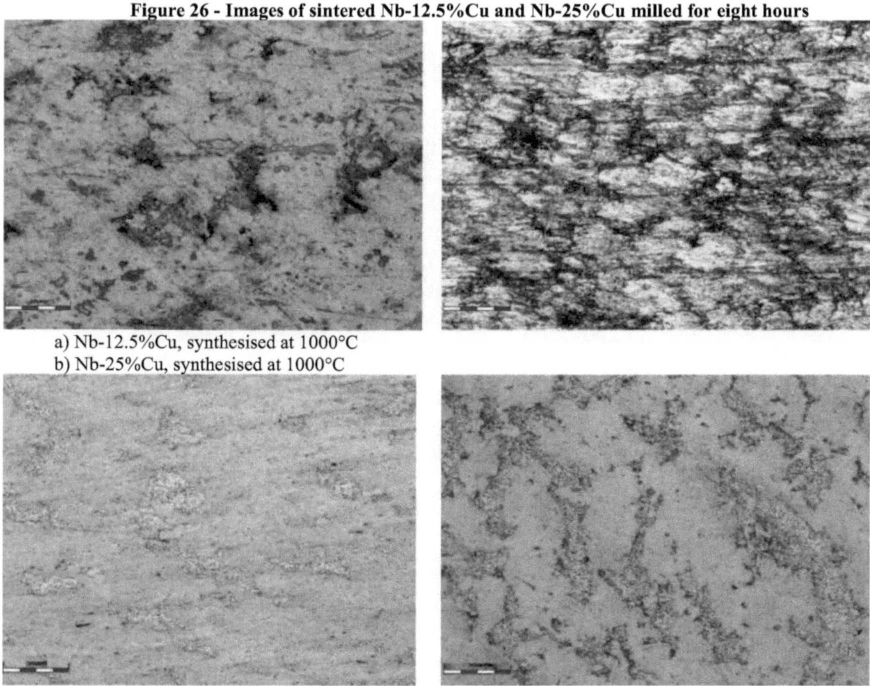

a) Nb-12.5%Cu, synthesised at 1000°C
b) Nb-25%Cu, synthesised at 1000°C

c) Nb12.5%Cu, sintered at 1100°C
d) Nb-25%Cu, sintered at 1100°C
Source: own elaboration

The sintered specimens were then imaged using a scanning electron microscope (SEM). Figure 27 shows the specimens sintered at 1000°C.

Figure 27 - Images of the specimens sintered at 1000°C

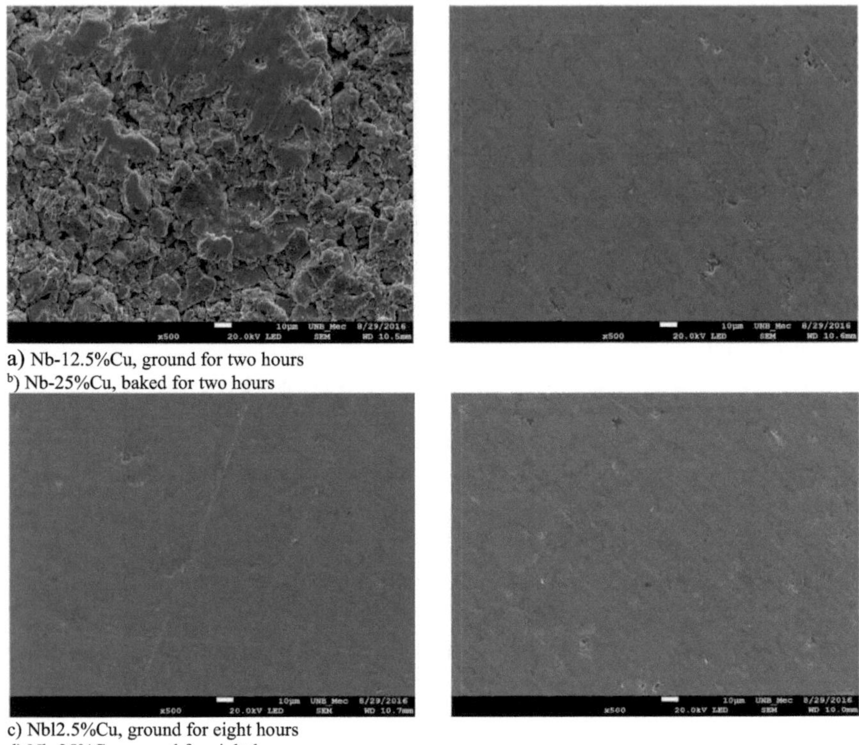

a) Nb-12.5%Cu, ground for two hours
^b) Nb-25%Cu, baked for two hours

c) Nb12.5%Cu, ground for eight hours
d) Nb-25%Cu, ground for eight hours
Source: own elaboration

Figure 27 shows that the hard-milled Nb-12.5%Cu composition had regions with large amounts of pores. Regions with pores were also observed in the sintered specimens "c" and "d" in figure 26. In view of this, it was realised that the increase in grinding time caused the regions with pores to decrease. Figure 28, which shows the specimens sintered at 1100°C, shows that the Nb-25%Cu specimen milled for two hours had regions containing pores. Synthesising at 1100°C produces a less porous composite material. Of the four specimens analysed, one showed a region with pores.

Figure 28 - Images of the specimens sintered at 1100 C

a) Nb-12.5%Cu, ground for two hours
b) Nb-25%Cu, baked for two hours

c) Nb12.5%Cu, ground for eight hours
d) Nb-25%Cu, ground for eight hours

Source: own elaboration

4.3 DENSITY

After milling the mixtures of niobium and copper powders, the mass of the powder compressed by knocking the container against the tabletop and its volume were measured. The apparent density of the alloys was then calculated. Table 3 shows these results.

Table 3: Apparent density of powder mixtures

COMPOSITION OF THE Nb-Cu POWDER MIX / MILLING TIME	APPARENT DENSITY (g/cm3)
Nb - 12.5%Cu / 2 (two) hours	4,07
Nb - 25%Cu / 2 (two) hours	3,89
Nb - 12.5%Cu / 8 (eight) hours	3,51
Nb - 25%Cu / 8 (eight) hours	3,26

Source: own elaboration

Analysing the apparent density, it was observed that the density decreases as the grinding time increases. This decrease is due to the flattening of the particles. As the metal powders collide with the balls and the walls of the grinding jug, plastic deformation occurs and, consequently, the particles flatten out. This causes the particle volume to increase. Figure 29 shows this behaviour

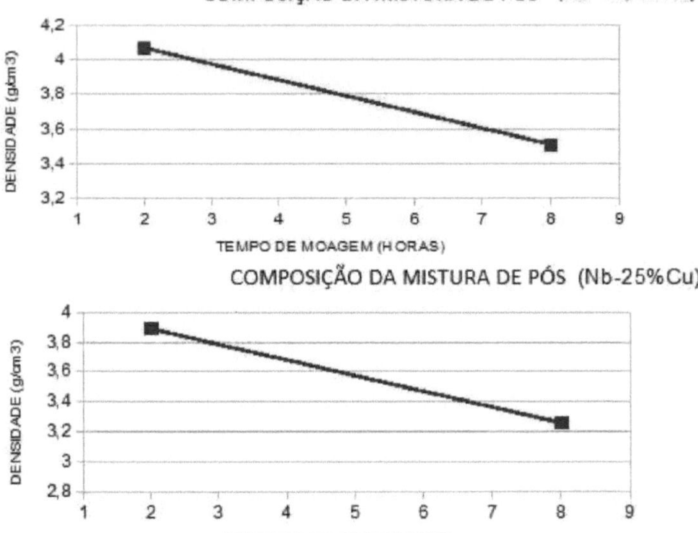

Source: own elaboration

After compacting the specimens, the diameter of the specimens and their respective height were determined. The density of the green compacted material was then determined using the geometric method. Table 4 shows these results.

Table 4: Density of the specimens before sintering

SPECIMEN COMPOSITION / GRINDING TIME	BATCH 1 TEST BODIES DENSITY (g/cm $)^3$	TEST BODY DENSITY OF BATCH 2 (g/cm $)^3$
Nb - 12.5%Cu / 2 (two) hours	5,77	5,79
Nb - 25%Cu / 2 (two) hours	5,88	5,84
Nb - 12.5%Cu / 8 (eight) hours	5,81	5,83
Nb - 25%Cu / 8 (eight) hours	5,90	5,87

Source: own elaboration

It was generally expected that higher density values would be seen in the samples milled with the shortest milling time, since a lower degree of hardening makes the composite particles less hard, making them more compressible.

Table 4 and figure 30 show that the density values of the green specimens are generally increasing when analysing samples with the same composition. One possibility for this is that grinding causes the particles to shrink. As the particles get smaller, they agglomerate more easily.

During compaction, smaller particles occupy spaces between the larger particles, and a process of accommodation takes place in the existing spaces. It was also observed that the densities of the green compacts increased as the composition of the copper in the composite increased.

Figura 30:   Density of green compacted materials

Samples from batch 1

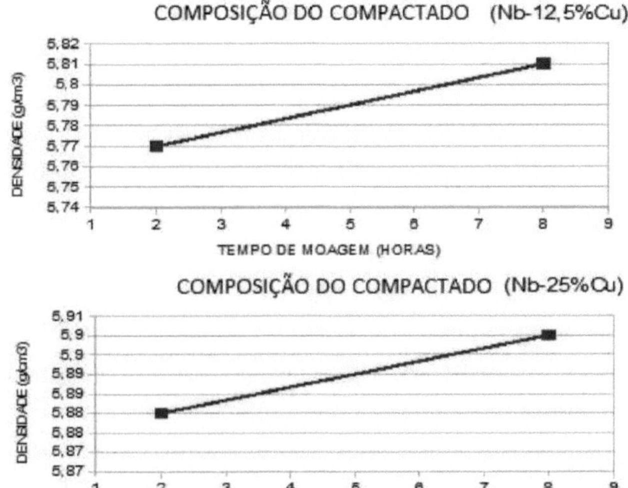

Samples from batch 2

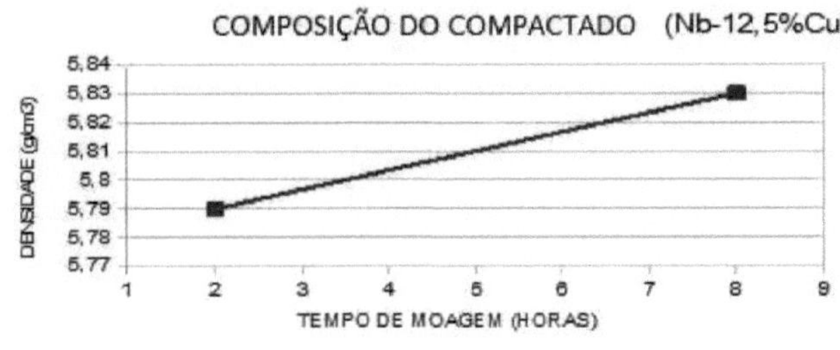

Source: own elaboration

After sintering batches 1 and 2 at temperatures of 1000°C and 1100°C respectively, the density of the specimens was calculated using the Archimedean method. The values are shown in the table below.

Table 5: Density of the specimens after sintering

SPECIMEN COMPOSITION / GRINDING TIME	BATCH 1 TEST BODIES DENSITY $(g/cm)^3$	TEST BODY DENSITY OF BATCH 2 $(g/cm)^3$
Nb - 12.5%Cu / 2 (two) hours	5,92	6,51
Nb - 25%Cu / 2 (two) hours	6,18	6,89
Nb - 12.5%Cu / 8 (eight) hours	6,26	6,54
Nb - 25%Cu / 8 (eight) hours	6,32	6,92

Source: own elaboration

From the information in figure 31, it can be seen that the samples from Batch 1, which were sintered at 1000°C, had the expected behaviour for density values. Materials ground for a longer period tend to increase in density. One of the possibilities after sintering is a decrease in the number of pores in the samples, causing the density to increase. Figure 31 shows this behaviour.

The graphs for batch 2 samples, which were sintered at 1100°C, also show an increase in the density of the samples. As the temperature rises, there is an increase in density, accompanied by spheroidisation and progressive closure of the voids. And through diffusion at the grain boundaries, the last rounded and isolated voids disappear.

In this way, the respective densities of the samples were calculated. From the values in Table 6, it can be seen that the densification increases for both longer grinding times and higher temperatures. Figure 31 shows this behaviour.

Table 6: Densification of the sintered specimens

SPECIMEN COMPOSITION / GRINDING TIME	DENSIFICATION OF BATCH 1 SPECIMENS (%)	DENSIFICATION OF BATCH 2 SPECIMENS (%)
Nb - 12.5%Cu / 2 (two) hours	69,23	76,13
Nb - 25%Cu / 2 (two) hours	71,83	80,08
Nb - 12.5%Cu / 8 (eight) hours	73,20	76,48
Nb - 25%Cu / 8 (eight) hours	73,46	, 80,43

Source: own elaboration

Figure 31: Density of sintered samples

Samples sintered at 1000°C (batch 1)

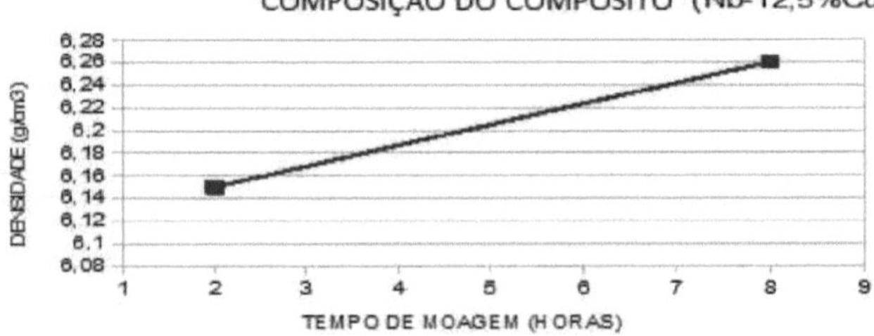

Samples sintered at 1100°C (batch 2)

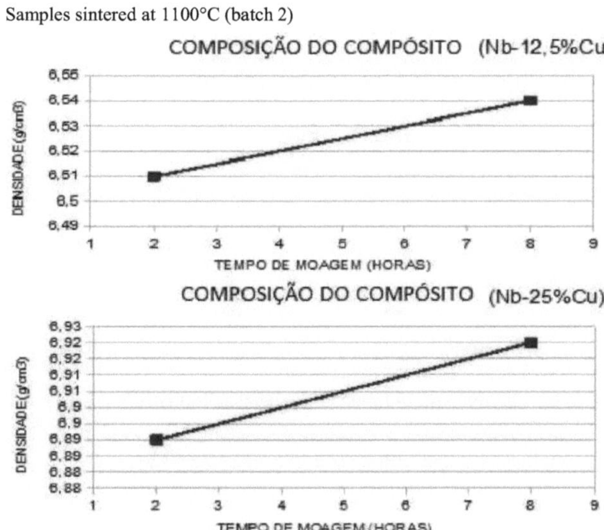

Source: own elaboration

1.4 COMPOSITION OF THE SPECIMENS

Qualitative and semi-quantitative analysis of the powders and sintered bodies was carried out using the EDS detector installed inside the microscope. Five regions were chosen for each sample. The average of the actual compositions and the standard deviation were calculated. Table 7 was constructed from this analysis.

Table 7: EDS analysis of Nb-Cu composite powders

SAMPLES AND THEIR THEORETICAL COMPOSITION	ACTUAL COMPOSITION
Nb-12.5%Cu ground for two hours	Cu =12.3 ±0.7% Nb = 84.522 ± 0.6 % Zn = 3.1782 ±0.3 %
Nb-25%Cu ground for two hours	Cu = 23.84±3% Nb = 72.928 ± 2.8 %

43

	Zn = 3.232 ±0.3%
Nb-12.5%Cu ground for eight hours	Cu =13.74 ±0.7% Nb = 83.788 ± 0.8 % Zn = 3.138 ±0.1 %
Nb-25%Cu ground for eight hours	Cu = 26.086 ±1.4% Nb = 70.816 ± 1.4% Zn = 3.098 ±0.1 %

Source: own elaboration

In all the composite samples produced, it can be seen that the actual composition is close to the theoretical composition. The presence of zinc is due to the insertion of zinc stearate in the samples. The purpose of zinc stearate is to lubricate the mixture of composite powders, preventing the powders from adhering to the grinding balls and jar. Another function of zinc stearate is to make it easier to remove the pressed part from the compacting die

EDS analysis was also carried out on the sintered samples. The procedure adopted consisted of choosing five regions from each sample. The measurements of the actual compositions were used to calculate the mean and standard deviation. The results are shown in Tables 8 and 9.

Table 8: EPS analysis of the Nb-Cu composite sintered at 1000°C

SAMPLES AND THEIR THEORETICAL COMPOSITION	ACTUAL COMPOSITION
Nb-12.5%Cu ground for two hours	Cu = 13.496 ±1.4% Nb = 86.454 ± 1.4 % Zn = 0.05 ± 0.1 %
Nb-25%Cu ground for two hours	Cu = 22.316 ±4.2% Nb = 77.618 ± 4.2 % Zn = 0.066 ±0.1 %
Nb-12.5%Cu ground for eight hours	Cu =11.092 ±2.7% Nb = 88.874 ± 2.7 % Zn = 0.034 ±0.1 %
Nb-25%Cu ground for eight hours	Cu = 23.314 ±5.6% Nb = 76.582 ± 5.6 % Zn = 0.104 ±0.1 %

Source: own elaboration

Table 9: EPS analysis of the Nb-Cu composite sintered at 1100°C

SAMPLES AND THEIR THEORETICAL COMPOSITION	ACTUAL COMPOSITION
Nb-12.5%Cu ground for two hours	Cu =13.818 ±1.8%

	Nb = 86.124 ± 1.9%
	Zn = 0.058 ±0.1 %
Nb-25%Cu ground for two hours	Cu = 22.04 ±5.6%
	Nb = 77.874 ± 5.6 %
	Zn = 0.086 ±0.1 %
Nb-12.5%Cu ground for eight hours	Cu =12.218 ±1.4%
	Nb = 87.748 ± 1.4 %
	Zn = 0.034 ±0.1 %
Nb-25%Cu ground for eight hours	Cu = 23.308 ±3.6%
	Nb = 76.566 ± 3.6 %
	Zn = 0.126 ±0.1 %

Source: own elaboration

Tables 8 and 9 show that a small amount of zinc stearate remained in the samples. Maintaining a temperature of 450 °C for one hour in the two sintering processes (at 1000 °C and 1100 °C) did not completely remove the zinc stearate.

In all the samples, there was variation in the real composition compared to the theoretical composition, since the results are obtained from the point chosen in a sample. Therefore, the point chosen for this chemical composition probably had a greater or lesser amount of one of the components. In this way, because EDS analysis is carried out on a specific region, it implies a variation in the percentage of its components.

1.5 CHARACTERISATION OF THE CRYSTAL STRUCTURE

After carrying out X-ray diffraction of the sintered samples, using the Cu element Ka as the radiation source (30 kV and 30 mA), with an angular range (in 20) of 10° to 80° at a scanning speed of 2°/min, Figures 32 and 33 were obtained. These figures show the diffractograms.

Figura 32:   Diffractograms of samples sintered at 1000°C

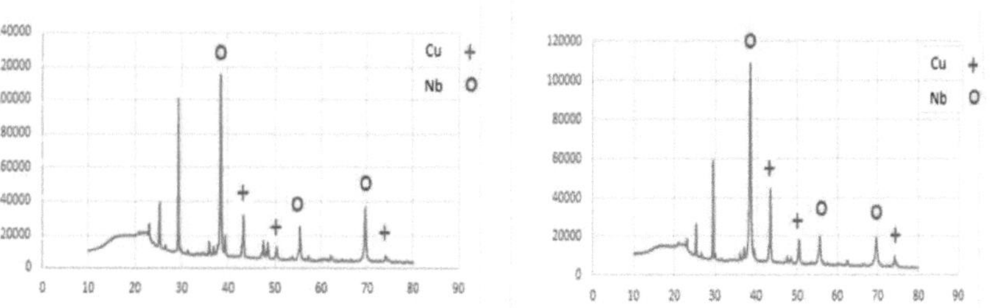

a)Diffractogram - Nb-12.5%Cu composite (ground for 2 hours and sintered at 1000°C)
b)Diffractogram - Nb-25%Cu composite (ground for 2 hours and sintered at 1000°C)

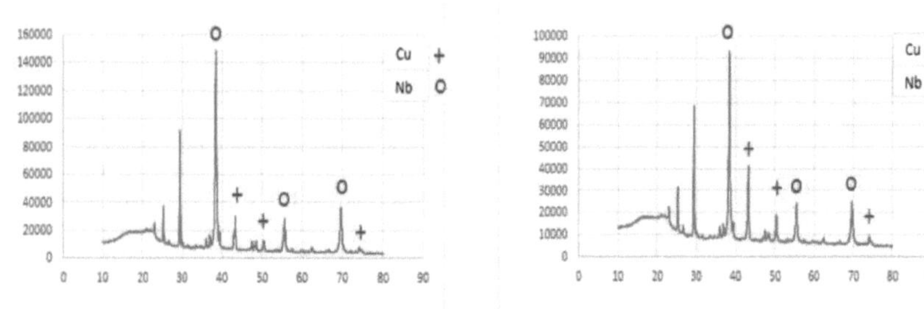

c)Diffractogram - Nb-12.5%Cu composite (ground for 8 hours and sintered at 1000°C)
d)Diffractogram - Nb-25%Cu composite (ground for 8 hours and sintered at 1000°C)

Source: own elaboration

Analysing figure 32, we can see that the increase in milling time did not cause the elements that make up the composite to amortise. It is known that the processing of powders by high-energy milling involves the events of deformation and fracture of the particles. These events produce defects in the crystalline network and grain refinement as the milling time increases. This favours the formation of amorphous phases.

The diffractograms of the samples sintered at 1100°C were also constructed. These diffractograms are shown in Figure 33.

Figura 33: Diffractogram of samples sintered at 1100°C

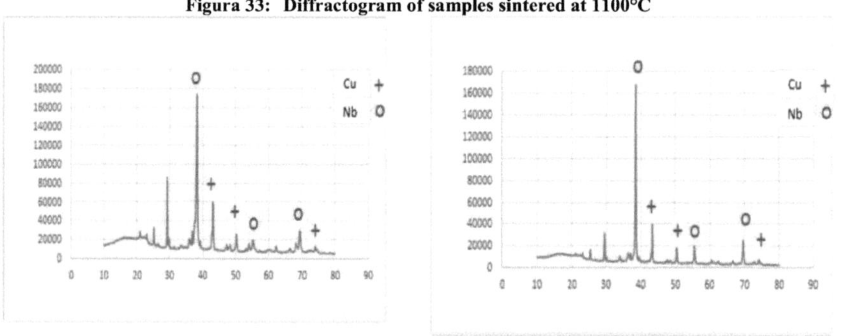

Diffractogram - Nb-12.5%Cu composite (ground for 2 hours and sintered at 1100°C)
Diffractogram - Nb-25%Cu composite (ground for 2 hours and sintered at 1100°C)

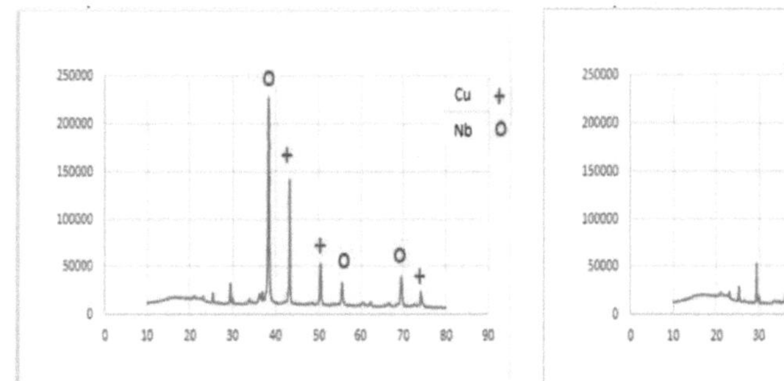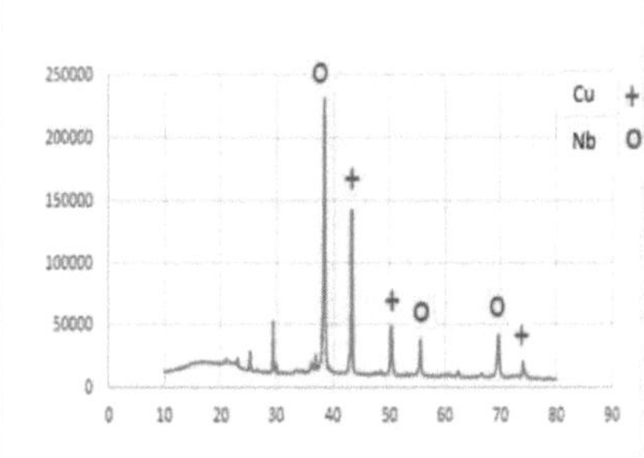

Diffractogram - Nb-12.5%Cu composite (ground for 8 hours and sintered at 1100°C)
Diffractogram - Nb-25%Cu composite (ground for 8 hours and sintered at 1100°C)
Source: own elaboration

All the diffractograms show the presence of a high-intensity peak between 37 and 40° corresponding to the presence of niobium in the sample's microstructure, predominantly in the (011) and (020) crystallographic planes, as well as the angles at 56° (corresponding to the 002 plane) and close to 70° (220). The peak corresponding to the presence of copper showed low intensity, between 42 and 45°.

There were also peaks of calcite and calcium oxides of considerable intensity between angles 28 and 32°. One possibility for this is contamination by the material that makes up the crucible in which the samples are placed for sintering.

A thicker peak was seen between angles 10 and 22. This behaviour is due to the presence of phenolic hot-melt resin. The Nb-Cu composite samples were embedded with this type of resin and taken to XRD analysis.

4.6 VICKERS MICROHARDNESS

The mechanical Vickers microhardness test was carried out by applying a force equal to 1000 gF over a time interval of 10 seconds. Table 10 and figure 34 were obtained.

Table 10: Vickers microhardness of sintered specimens

SPECIMEN COMPOSITION / GRINDING TIME	VICKERS MICROHARDNESS OF BODIES SINTERED AT 1000°C (HV)	VICKERS MICROHARDNESS OF BODIES SINTERED AT 1100°C (HV)
Nb - 12.5%Cu / 2 (two) hours	21,66±1,9	27,56±0,8
Nb - 25%Cu / 2 (two) hours	28,98±2,5	29,86±1,3
Nb - 12.5%Cu / 8 (eight) hours	29,50±0,7	31,18±0,5
Nb - 25%Cu / 8 (eight) hours	30,40±1,6	34,80±0,5

Source: own elaboration

From the information in Table 10, it can be seen that the microhardness increased for materials ground for a longer period of time. This behaviour is to be expected, since during compaction the particles of the composite material undergo plastic deformation. As the grinding time is increased, the hardening process takes place. This causes the particles to become harder. Figure 34 shows this behaviour.

Figura 33:   Vickers microhardness of the sintered samples

Samples sintered at 1000°C (batch 1)

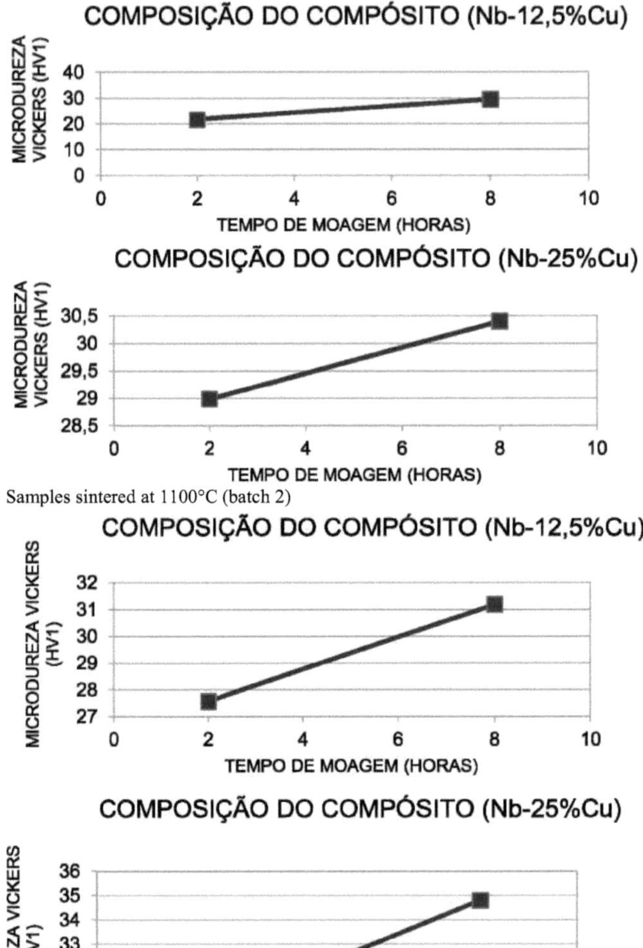

Samples sintered at 1100°C (batch 2)

Source: own elaboration

It can also be seen that temperature influenced hardness, with samples sintered at a higher

48

temperature having higher hardness values for the same grinding conditions.

CHAPTER 5

CONCLUSIONS

The general objective of this work was achieved, since the Nb-Cu composites were obtained via the powder metallurgy process, in the proposed compositions of 12.5% and 25% Cu by mass. The tests carried out made it possible to characterise the samples and reach some conclusions:

a)	Morphology: The niobium powders have an angular morphology due to their processing using the HDH technique. The copper powders have a spherical and porous morphology because they are processed using the carbonyl process. After mixing and milling the elemental copper and niobium powders, it was concluded that for longer milling times there is an improvement in the distribution of the Nb-Cu composite particles, i.e. they make the material more homogeneous. As the milling time increases, it is expected that there will be a greater degree of amortisation of the elements that make up the Nb-Cu composite. The collisions of the niobium and copper particles with the grinding bodies cause changes in the crystalline structure.

b)	Density: The apparent density of powder mixtures decreases with increasing milling time. As the milling time increases, the shocks of the powders against the milling bodies cause the particles to flatten out. This leads to an increase in the volume of the powder mixtures. It was also concluded that the density of the green compact tends to increase with very long milling times, since they make the particles smaller and, consequently, facilitate agglomeration. This leads to a higher green density. The densities of sintered specimens increase at higher temperatures. As the temperature increases, there is a progressive

closing the voids through diffusion at the grain boundaries. This can reduce pores and increase density.

c)	Influence of the temperature factor: Temperature influences sintering. Higher temperatures result in higher quality sintering.

d)	Microhardness: Longer milling times and sintering at higher temperatures tend to increase microhardness values.

5.1 SUGGESTIONS FOR FUTURE WORK

Below are some suggestions for future work:

- Increase the rotation speed of the grinding process;
- Increase the sintering temperature;

- Analysing electrical and thermal conductivity;
- Increase the grinding time.

REFERENCES

ANDRADE, M. L. A.; et al. Copper industry - industrial operations area. **BNDES,** 1997.

ASM, I. Materials handbook. Vol. 7. **Powder Metal Technology and Applications, Materials Park,** 1990.

BAKER, H. Metals handbook. Introduction to alloy phase diagrams. **ASM International,** USA, vol. 3, 1992.

BRITO, F. L; et al. A theoretical study on sintering in powder metallurgy. **Holos,** vol. 3, p. 204-211,2007.

CERNIAK, S. N. **Study and development of a niobium electrolytic capacitor.** Natal, 2011, 120 p. Master's dissertation. Federal University of Rio Grande do Norte.

CHIAVERINI, V. **Powder metallurgy.** São Paulo: Brazilian Association of Metallurgy and Materials, 2001.

. **Mechanical technology - mechanical construction materials.** São Paulo: Makron Books, 1986.

COSTA, F. A.; et al. Composite Ta-Cu powder prepared by high energy milling. **Int. J. Refractory Metals & Hard Materials,** vol. 26, p. 499-503, 2008.

; AMBROZIO FILHO, F.; SILVA, A.G.P. Sintering behaviour in solid State of a W- 25%wtCu composite powder prepared by mechanical alloying. In: Powder metallurgy world congress, October 17-21, 2004, Vienna. **Proceedings of the powder metallurgy world congress,** Vienna: EPMA. 2004.

DELFORGE, D. Y. M.; et al. Sintering a mixture of stainless steel chips and stainless steel powder. A new technology for metal recycling? Rem: **Rev. Esc. Minas,** vol. 60, n.l, Ouro Preto, Jan./Mar. 2007.

FIGUEIREDO, N. C. **Metallurgical characterisation of phases in an Fe-30Ni alloy processed by**

powder metallurgy. Monograph (Obtaining the degree of Metallurgical Engineer), Federal University of Ceará, Fortaleza, 2013.

FOGGIATTO, B.; LIMA, J. R. B. Macroeconomic analysis of the main Brazilian mineral assets. **PIC - EPUSP,** n. 2, 2004.

FROES, F.H.; EYLON,D. Production of titanium powder. Metals handbook. 9 th ed.,1984, p 164-167.

GERMAN, R. M.; BOSE, A. Injection moulding of metals and ceramics. **Metal Powder Industries Federation,** 1997.

GOMES, U. U. **Tecnologia dos pó: fundamentos e aplicações, Natal/RN.** Federal University of Rio Grande do Norte. University Publishing House, 1995.

POWDER METALLURGY SECTOR GROUP. Available at: <www.metalurgiadopo.com.br>. Accessed on: 18 August 2016.

JÚNIOR, N. V.; USE OF THE MASTER ALLOY POWDER ALLOY TECHNIQUE IN THE OBTAINING OF COMPONENTS IN 4140HC STEEL VIA THE INJECTION POWDER MOLDING PROCESS, Graduation, Federal University of Santa Catarina, Florianópolis, 2007.

KNEWITZ, F. L. **Comparative study of NiTi samples produced by powder metallurgy.** Porto Alegre: UFRGS. 2009.

LEE, W. E.; REINFORTH, W. M. **Ceramic microstructures property control by Processing.** In: LEE, W. E.; REINFORTH, W. M. London: Chapman & Hall, 1994. p. 03-65.

LENEL, F. V.; et al. Powder metallurgy principies and applications. Princeton, New Jersey: **Metal Powder Ind. Federation,** ch. 9, 10 and 11, p. 241-319, 1980.

LIMA, H. de. **Study of a sintered Nb-15%pCu obtained by high-energy milling and liquid-phase sintering.** Dissertation (Master's Degree), Federal University of Rio Grande do Norte (UFRN), Natal-RN, 2015.

MAURICE, D. R.; COURTNEY, T . H. Modelling of mechanical alloying: part III. Applications of computational programmes. **Metallurgical and Materials Transactions A,** vol. 26A, n. 9, p. 2437-44, Sept. 1995.

MELCHIORS, G. **Characterisation of Nb-20%cu composites obtained by high-energy milling and liquid-phase sintering.** Dissertation (Master's Degree) - UFRN, Natal, 2011.

. **Obtaining and characterising Nb-Cu composites obtained by high-energy milling and sintered in a vacuum furnace.** VI National Congress of Mechanical Engineering. 18 to 21 August 2010, Campina Grande-PB, Brazil, 2010.

MEYERS, M. A.; MISHRA, A.; BENSON, D. J., Mechanical properties of nanocrystalline materials. **Progress in Materials Science,** Univ. of California, USA, 2005.

MITKOV, M.; BOZIC, D. Hydride-Dehydride conversion of solid TÍ6A14V to poder form. Mat. Charact. n. 37, 1996, p 53-60.

MORO, N.; AURAS, A. P. **Manufacturing processes - powder metallurgy and the future of industry.** Florianópolis: Federal Technological Education Centre of Santa Catarina, 2007.

NIKOLAEV, A. K.; ROZENBERG, V. M. Heat Resistant Steels And Alloys. Properties of Cu-Nb alloys. **Metal Science and Heat Treatment,** vol. 14, Issue 10, p. 888-890, Oct 1972.

ROTTA, M. **Nb-Cu composite obtained by mechanical action and sintering - physical, thermal and electrical behaviour.** Dissertation (Master's Degree) - State University of Maringá, Maringá-PR, 2005.

SALGADO, L. **Processing of iron-nickel-copper-molybdenum alloy by high-energy grinding.** Thesis (Doctor of Science in Nuclear Technology - Applications), IPEN, University of São Paulo, São Paulo, 2002.

SCHAERER, M. M. **Numerical analysis of the densification behaviour of metallic powders resulting from the uniaxial compaction process.** PhD thesis, Rio de Janeiro, 2006.

SILVA, A. G.; JÚNIOR, C. A. Rapid sintering: its application, analysis and relation with the innovative sintering techniques. **Cerâmica,** p. 17, 1998.

SURYANARAYANA, C. Mechanical alloying. In: ASM Handbook. (Ed.) Powder metal technologies and applications. **Materials Park,** OH, vol. 7, p. 80-179, 1998.

TORRALBA, J.M.; COSTA, C.E.; VELASCO, F., P/M aluminium matrix composites: an overview. **Journal of processing technology,** A 133, p. 203-206, 2003.

More
Books!

info@omniscriptum.com
www.omniscriptum.com
OMNIScriptum

Printed by Books on Demand GmbH, Norderstedt / Germany

Obtaining the Nb-Cu Composite

The Nb-Cu composite can be used in the manufacture of electrical contacts, resistors, welding electrodes and more. These materials must have high electrical and thermal conductivities and heat resistance. The production of these composites using conventional technology is limited, since they involve chemical elements that have very different melting points. An alternative is to use powder metallurgy. The aim of this work is to obtain an Nb-Cu composite via powder metallurgy.

ACADEMIC TRAINING: Master's Degree in Integrity of Engineering Materials. University of Brasília, UnB, Brasília, Brazil Year completed: 2016. PROFESSIONAL ACTIVITY: Company: SEE / DF, Title: PUBLIC SERVANT, Position: INTERMEDIATE COORDINATOR. Company: SEDUCE / GO, Relationship: PUBLIC SERVANT, Position: TEACHER.

Semiconductor Nanostructures:

Excitons and Sorting for Optoelectronic Applications

Mohamed Chnafi
Omar Mommadi
Abdelaziz El Moussaouy